A PHOTOGRAPHIC GUIDE TO

ALPINE PLANTS
OF
NEW ZEALAND

Laurie Metcalf

T0307143

WHITE CLOUD BOOKS

Oh! I sigh for the mountain breezes,
The scents of fragrant flowers:
The glories of the landscape,
The joy of careless hours
Spent in the alpine gardens,
From care and worry free;
Oh! I long you too may know them,
And share this joy with me.

– Anon.

This print published in 2023 by White Cloud Books, an imprint of
Upstart Press Ltd.
First published in 2006 by New Holland Publishers (NZ) Ltd

Upstart Press Ltd.
26 Greenpark Road, Penrose, Auckland 1061, New Zealand

www.upstartpress.co.nz

ISBN: 978-1-77694-040-0

A catalogue record for this book is available at the National Library of
New Zealand

Printed by Dongguan P&C Printing Technology Co., China

10 9 8 7 6 5 4 3 2 1

Front cover photograph: North Island edelweiss (*Leucogenes
leontopodium*)
Back cover photograph: *Ranunculus haastii*
Spine photograph: Golden Spaniard (*Aciphylla aurea*)
Title page photograph: *Hebe macrantha*

Contents

Introduction

With a mountain chain extending over most of the length of the South Island and approximately two-thirds of the North Island, New Zealand is very much a hilly and mountainous land. It is not surprising, therefore, that the country is well endowed with a rich and varied alpine vegetation.

As might be expected, the majority of our native alpine plants are to be found in the more mountainous South Island, although a good representation of species also occurs in the North Island. For the native plant enthusiast a visit to the alpine zone of any of the main mountain regions can be very interesting because of the differing patterns of vegetation that are part of the changes in altitude.

What is an alpine plant? Generally, alpine plants are regarded as those that grow in or are confined to the alpine zone. The alpine zone may be defined as that area of the mountainous landscape that lies above the tree line. The tree line generally forms a sharp and distinctive, more or less horizontal line along the mountain slopes, where the growth of forest ceases. Below the tree line the forest colours everything a dark green while, above it, the landscape appears, from a distance, to be a tawny shade. That is because of the presence of the tussock grasses that typify much of the alpine vegetation.

Many people imagine that to qualify as an 'alpine' a plant must be small, maybe no more than 15 to 25 cm in height, and very compact in its growth. In truth, any plant that inhabits the alpine zone is an 'alpine' regardless of whether it grows no more than 1 cm tall or reaches more than 1.5 m tall.

Alpine plants are many and varied, belonging to a whole range of plant families as well as presenting themselves in all shapes and sizes. Some are highly specialised so that they are able to survive only in particular habitat conditions. Those that inhabit the mountain shingle screes are a case in point. There are buttercups (Ranunculaceae) that range from giants up to a metre tall to midgets no more than a couple of centimetres in height. As well as having species with typical daisy-type flower heads, the daisy family (Asteraceae) has a number of highly specialised species that grow only on very exposed high-altitude rocks and bluffs.

One of the interesting features of our alpine flora is the lack of brightly coloured flowers when compared with the floras of other countries. The majority of New Zealand's alpine flowers are white or yellow. It is only when you examine them more closely that you realise that a number do have traces of other colours in them. It is just that such colours exist only as veining, spots or ocular (eye-like) rings and so are not always obvious to the casual observer. Some of the *Ourisia* species may exhibit a pinkish flush on the reverse sides of their flowers, while others have violet in the throats of the corollas. *Parahebe* species, a number of *Myosotis* and some species of *Gentianella* may all display some sign of colour in their flowers. It is considered that brightly coloured flowers probably offered no advantage to alpine plants, particularly with regard to pollination.

In New Zealand we appear to have a lack of specialised pollinating insects, which, together with the fact that mountain areas are frequently quite prone to winds, means that whatever insects inhabit these regions are not always encouraged to fly during the day. Some day-flying moths and the occasional butterfly may be seen but, generally, pollinating insects are not overly conspicuous. As any visitor to the mountain areas will know, although breezy conditions may be frequent during the day, the wind generally drops towards evening and calmer conditions will then prevail. It is during this late afternoon and early evening that many pollinating insects are on the wing, particularly moths. In the dimmer light of the evening, white flowers are more easily visible to pollinating insects. Quite a number of alpine species also have scented flowers which, no doubt, act as an attractant to some pollinators. In any case, the real reasons for the seeming lack of colour in the alpine vegetation are still not completely understood.

New Zealand's native flora comprises almost 2500 different species of plants. Of those, some 600 or more species may be regarded as true alpines because they are confined to the alpine zone. Additionally, a further 350 or more species occur in both lowland and mountain areas so that our high-mountain flora comprises almost 50 per cent of the total flora. Of that flora, at least 420 to 450 species of high-mountain plants are confined to the mountains of the South Island. It has been said that in New Zealand's alpine regions there is a greater range of alpine vegetation than occurs in most other parts of the world.

How to use this book

Headings

With each listing, the common name (where applicable) and scientific name are given at the top of the page. The family that a species belongs to is given in the coloured tab at the side of the page. For ease of reference, the species described within a family are listed in alphabetical order, according to their scientific names, so that their affinities with related species may easily be seen. The scientific names are those in current use, although more recent research may see some of those names superseded.

Wherever possible, common names are provided but, with many alpine plants, there are just no common names in use. At various times, well-meaning authors have attempted to provide or coin common names for all of the plants that they have mentioned, usually with pretentious or affected results. Also, not all plants are sufficiently notable for them to have been given common names. As more people frequent the mountains, and take note of the plants about them, then no doubt more common names will eventuate.

Terminology

Every attempt has been made to keep all descriptions as straightforward as possible and to keep botanical terminology to a minimum, but it is not always possible to avoid the use of some technical terms.

Sometimes their use enables greater accuracy and at others it can avoid the use of wordy phrases that may be better expressed by just the one word. A glossary of the terms used has been provided on p. 126.

Distribution maps

Distribution maps are provided in order to give an indication of the broader areas (in green shading) over which a particular species is likely to be present. These maps do not necessarily signify that any particular species will occur in every part of the area indicated. Some species have remarkably discontinuous distributions while, depending upon the part of the country, there may be considerable distances between one sighting of a particular species and the next location where it may appear. Other species have quite wide distributions and may occur frequently over large parts of the indicated areas.

The alpine flora of New Zealand

After leaving the zone of lowland vegetation, the next zone is known as the montane zone. It literally means 'of or inhabiting the mountains'. The montane zone may encompass forest, scrublands and grasslands or, sometimes, combinations of those three vegetation types. While the montane zone may consist of high and steep hills and mountains, it is only when the tree line is reached, and passed, that alpine conditions may be said to prevail. Apart from the vegetation, alpine conditions are, among other factors, also influenced by snowfall and the length of time that snow lies on the ground during each year.

As already stated, the alpine zone commences at the upper limit of the tree line and, generally, it extends for about the next 1000 m or so of vertical altitude. The tree line actually varies considerably from north to south, being highest, as might be expected, in the warmer North Island and it progressively attains its lowest altitude in the far south of the South Island. In the North Island the tree line commences at about 1500 m and by the time that it reaches Fiordland, in the far south, it is down to about 900 m.

Compared with northern-hemisphere countries, New Zealand has a very low tree line and, at the same time, a correspondingly low permanent snow line. While the altitude of the tree line progressively decreases from north to south there can also be local variations that appear to defy logic. For example, on the wet side of the main divide warmer conditions allow for a higher tree line, whereas on the colder and drier eastern side there may be local areas where the tree line is lower than the average for that area.

The alpine zone itself is divided into two zones or belts according to altitude and the kind of vegetation that is present. Firstly, at the lower level, just above the tree line, there is what is classified as low-alpine vegetation, and above that is the beginning of the high-alpine vegetation. (Beyond that you arrive at the region of

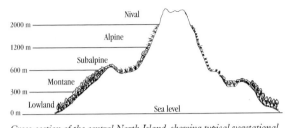

Cross-section of the central North Island, showing typical vegetational zones and altitudinal limits (not to scale).

bare rock and snow where virtually no plants grow, known as the nival zone.) Each of those two vegetation zones is divided into successive zones or bands of vegetation (often referred to as associations) that change according to altitude and possibly the aspect of the mountain slopes. For ease of reference it is easier to assume that these plant associations may change at approximately every 100 m gained in vertical altitude. Of course, that may vary somewhat, depending on the region.

A. Low-alpine vegetation
Within the low-alpine vegetation zone there are five bands or associations of plants. They are detailed below.

1. Mixed snow tussock–scrub
This is the zone that commences just above the tree line and continues upwards for approximately the first 100 m of altitude. It comprises mainly woody shrubs but also includes a number of the larger herbs. At least one species of *Chionochloa* (large snow tussock) may be present, while there may be one or more of the larger species of *Hebe*, such as *H. odora*, perhaps *Olearia arborescens*, usually a *Dracophyllum* (grass tree or turpentine scrub, possibly *D.*

Lewis Pass, North Canterbury, with mixed snow tussock–scrub. Mountain beech (Nothofagus solandri *var.* cliffortioides) *on the distant mountains.*

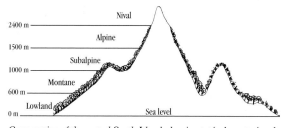

Cross-section of the central South Island, showing typical vegetational zones and altitudinal limits (not to scale).

uniflorum), and *Coprosma*, which may include *C. pseudocuneata* and possibly *C. serrulata*. *Podocarpus nivalis* (mountain totara) is nearly always present as is *Phyllocladus alpinus* var. *aspleniifolius* (mountain toatoa). Herbs such as *Celmisia semicordata*, possibly *Ranunculus lyallii* (mountain lily) and maybe one of the larger species of *Aciphylla* (speargrass) will complete the picture.

2. Snow tussock–herbfield
From the mixed snow tussock–scrub association there is a gradual merging with the snow tussock–herbfield. This association mainly occurs in the wetter mountain regions but it may also occur on some of the eastern ranges where conditions are not too dry. Again, snow tussock–herbfield may include one species or another of the larger *Chionochloa* tussocks (*C. flavescens* is often the main species), some of the larger mountain daisies such as *Celmisia semicordata* and one or two of the smaller species, *Ranunculus lyallii* (mountain lily), as well as one of the smaller species of buttercup, and *Aciphylla* (speargrass). *Dracophyllum uniflorum* may be scattered among the snow grass tussocks and among the scattering of medium to smaller shrubs may be the large silvery tussocks of *Astelia nervosa*, while *Ourisia* and one or more species of *Hebe* can be conspicuous.

3. Herbfield
Rising another 100 m or so from the snow tussock–herbfield is the low-alpine herbfield. Generally, this is characterised by quite a variety of small and large herbs. They may include *Celmisia* (mountain daisies), *Aciphylla* (speargrass), *Ranunculus* (buttercup), *Astelia*, *Anisotome* and *Ourisia*. Depending upon the area there may be a small species of *Acaena* (biddy biddy), as well as *Gaultheria* and *Euphrasia* (eyebright). In some areas *Chionochloa australis* (carpet grass) may also appear.

4. Snow tussock grassland
This zone or association occurs mainly on the drier eastern ranges of the South Island rather than the wetter ranges about and west of the main divide. This grassland also occurs around the Volcanic Plateau of the North Island where *Chionochloa rubra* (red tussock) occurs. In the South Island the main species are *Chionochloa rigida*

(narrow-leaved snow tussock), *Chionochloa macra* (slim snow tussock) and, on some of the lower mountain country in Otago and Southland, *Chionochloa rubra* (red tussock).

Chionochloa rigida extends from the Rakaia River valley in Mid Canterbury southwards to northern and western Southland. From an ecological point of view, these tussock grasslands are very important. Where they are able to grow undisturbed their foliage canopies touch or almost touch, thus providing an almost complete cover for the ground, protecting it against excessive drying out, even under the worst of conditions. Their flowing foliage collects any moisture that may occur as fog or light rain and conducts much of it to the ground where it is protected by the foliage. The tussocks also protect the ground against the effects of the drying winds that can be so frequent in mountainous areas. Although some people see the tussock-clad landscape as uninteresting, our snow tussock grasslands give the landscape of South Island mountain areas an appearance that is so characteristic of New Zealand. This is especially the case when there is a strong westerly wind that causes whole tussock-clad hillsides to come alive as they wave and ripple with each succeeding wind gust.

Unfortunately, the requirements of pastoral grazing have not been good for much of our tussock grasslands. The tussocks have been weakened by excessive burning while, at times, over-grazing has exacerbated that effect. The result has been an invasion of undesirable plant species such as the dreaded *Hieracium* (hawkweed). Browsing by introduced mammals such as deer, rabbits and hares has not helped the situation. Where the snow tussock grasslands are left to their own devices and protected from harmful outside influences they will eventually recover.

5. Bogs

Alpine bogs usually occur in areas of very poor drainage. They differ from a swamp in that the whole surface of the ground is not covered by water, even though the soil is at all times saturated with water and shallow pools may be present. Usually the soil is soft and the water table is nearly always fairly close to the surface. In alpine regions bogs usually occur on level or nearly level ground or where there is a fairly gentle slope near the foot of a steeper slope. Sometimes, however, small boggy areas can occur on steeper slopes, but they should not be confused with seepages that can create boggy conditions, usually over smaller areas. They are more common in the higher rainfall regions and less so in the drier ranges east of the main divide. Such boggy areas are also referred to as herb-moor or cushion bogs.

Sometimes bright-green cushions may indicate the presence of sphagnum moss, either in greater quantity or perhaps sometimes as a more recent inhabitant of the bog. If the bog is large enough there may be some tussocks of *Chionochloa rubra* (red tussock) around its periphery or on higher parts within the bog. Apart from various sedges and rushes that normally inhabit bogs, other plant inhabitants may be *Celmisia alpina* (daisy), *Donatia novae-zelandiae, Drosera*

The Old Man Range, Central Otago, with cushion bog in the foreground and some late snow-bank in the distance.

(sundew), possibly one or two *Dracophyllum* species and *Coprosma perpusilla*. Also likely to be present are *Lepidothamnus laxifolius* (pygmy pine), *Pentachondra pumila* and *Phyllachne colensoi*.

B. High-alpine vegetation

At the upper limits of the low-alpine vegetation belt there is a gradual transition from that belt to the high-alpine vegetation belt. The plants become lower in stature, shrubs are more scattered as well as being smaller and more compact, while greater amounts of rock begin to show through the vegetation. This increasing amount of visible rock marks the beginning of the fellfield which is the first of the high-alpine vegetation associations.

6. Fellfield

The origin of this term comes from the Danish 'Fjeldmark', which translates into English as 'fell', meaning a barren or stony hill, and 'field' meaning a wide expanse; the combination of those two terms refers to the bare, stony ground that so typifies the New Zealand fellfield.

Even though there is often a considerable amount of bare ground showing between plants in the fellfield, it does not necessarily mean that the area is without interest; in fact, the reverse is often the case. In the fellfield it is possible to find some of the real gems of the alpine flora. They include members of the genus *Raoulia* and *Haastia* which, collectively, are known as 'vegetable sheep'. Some of the smaller species of speargrass may be observed along with smaller species of *Celmisia*, *Brachyglottis*, *Gentianella*, and some of the smaller species of *Hebe* as well as possibly *Chionohebe*.

There can be quite a difference between the fellfields in dry mountain regions and those of the wet mountain regions. In the latter there can be a much more interesting selection of species than may be found on the drier ranges down the eastern side of the country.

7. Scree

In some respects shingle screes may be regarded as being a kind of dry fellfield that mainly occurs on the drier ranges east of the main divide. As an alternative, screes are also known as talus slopes. The term 'scree' is derived from the old Norse 'skritha' and was originally applicable to any landslide on a hillside. In fact, in some parts of the South Island screes, or shingle screes, are still referred to as shingle slips.

Some screes are of comparatively recent origin, being no more than 100 to 150 years old. These screes date from the first European occupation of the high country, when sheep grazing and burning off of the protective plant cover allowed some of the steeper slopes to develop into the screes that are now regarded as being so typical of the drier South Island mountain ranges. More recent research, however, has shown that some more natural screes are much more ancient and predate the arrival of humans in this country. Generally, they are believed to be anything from 500 to 2500 years old. These older screes are generally much more stable than the induced screes that are of more recent origin.

The induced screes usually have a very loose surface layer that is quite mobile, so that when that layer of stones is disturbed it begins to move or run downhill, sometimes for quite a considerable distance. Beneath that loose and dry surface layer of stones there is usually a much firmer and more stable layer (about 10–20 cm below the surface) of finer material that always shows traces of moisture. In fact, at about 30–40 cm deep it may be quite moist to the stage of almost showing moving water. Many of the induced screes cover quite a wide altitudinal range, commencing almost at the snow line and terminating at the montane level.

Screes are remarkable plant habitats because of the challenging conditions they provide. During the summer the surface layer is quite dry and scorching hot, but with the advance of a sudden southerly front, within about 30 minutes the temperature may drop suddenly from about 54°C to 23°C. During spring and autumn, frost may be frequent so that any plant growth above ground may have to endure freezing during the night and sudden thawing as the sun rises during the morning. Early or late snowfalls may add to its problems.

During winter, conditions may not be quite so severe because while some scree plants are deciduous and die down for the winter, those that are evergreen are safely protected under a layer of snow.

Most of the plant species that inhabit screes are highly specialised. The colour of their foliage has adapted so that it is similar to that of the stone material of the scree. Their stems may be delicate, but they enable its top growth to move and not be damaged when the surface layer of the scree moves downhill. Alternatively, their stems may easily break off so that new ones can regrow to the surface. Their foliage is usually fleshy and thus able to resist heat and drought, while both surfaces of the foliage are protected by a thick skin or cuticle; their undersurfaces have the additional protection of a two-layered epidermis or cuticle, and they may also be further protected by a waxy bloom.

Generally, the screes appear to be bare of any living plants but a closer examination will show that a surprising number of plants may be scattered over the scree. There are some 25 species of plants that are more or less confined to shingle screes and some 10 or 12 of those are relatively common and may be observed on many screes.

8. Cushion vegetation
Cushion vegetation is generally more typical of the southern regions of Otago and Southland than the mountain ranges further to the north. It is a tundra-like association that consists of very dwarfed plants, usually those of a cushion-like habit, and is common on the rolling, plateau-like summits of the ranges of Central Otago and northern Southland. These ranges experience very strong winds, cool to cold summers and frosts at almost any time of the season. The soils may also be poorly drained so that it is a very difficult environment for the few species that are adapted to survive there. *Donatia novae-zelandiae* is one of the principal cushion species and in some areas may form extensive cushion bogs (also see 5. Bogs on p. 10). Other species may be *Anisotome imbricata*, *Phyllachne colensoi* and *P. rubra*.

9. Snow-bank vegetation
This is the highest association before the snow line and bare rock is attained. It is usually the area where snow from the previous winter may lie and accumulate, in hollows and large depressions that may be sheltered from the prevailing winds. In such areas the snow lies late and does not thaw until very late in the summer or, in some seasons, it may not thaw at all. The plants that live in snow-bank areas are able to exist through a very short growing season so that some, such as *Psychrophila*, are able to produce flowers right under the edge of the thawing snow. Other snow-bank plant species are *Celmisia haastii*, *Kelleria* species and *Chionochloa oreophila* (snow-patch grass).

Coral lichen *Cladia retipora*

The coral lichen is a very pretty species that is well named because it greatly resembles a piece of coral. In favourable conditions it can form cushions of quite a substantial size, reaching up to 1 m in diameter, although generally it is somewhat smaller. It is usually white or pale greyish but sometimes may be faintly yellowish and may often have a pinkish tinge. Its **vegetative parts** form a reticulated network and are very delicate and lace-like. When dry the coral lichen is very brittle but it can soon absorb any available moisture to become quite soft and spongy. The **branches** of coral lichen are 1.5–5 mm in diameter and they may be 2–5 cm or more in length. The branches are hollow and perforated, from base to apex, so that they have that coral-like appearance from which its common name is derived. Coral lichen occurs throughout New Zealand: in the North Island from Northland to Wellington, in the South Island from Nelson to Southland, on Stewart Island and the Chatham Islands and then extending southwards to the Auckland, Campbell and Antipodes islands. It is found on peaty soils often among tussocks and other low vegetation, especially in *Leptospermum* (manuka) and *Dracophyllum* heaths, in fellfields and sometimes on rocks, logs and sand-dunes. It ranges from sea level to 1200 m.

Aaron's beard *Usnea capillacea*

Aaron's beard is common in alpine regions, from the Volcanic Plateau southwards, where as a species of lichen it forms a prominent feature, particularly on *Nothofagus* or southern beech trees. It forms **beard-like growths** that are pendulous from the branches of the beech trees and hang down for 25–30 cm. It is grey-green in colour and the **slender stems** of its tufts are much entangled so that it is difficult to separate one from the other. This kind of lichen is known as a mist species, which indicates that the kind of areas where it grows are frequently shrouded in mist. The species of *Usnea* are very difficult to identify and sometimes there are intermediate forms so that it is not easy to determine whether a particular plant is a valid species or not. Other species of *Usnea* are sometimes common on bushes of small shrubs that grow above the tree line, such as *Dracophyllum* and other subalpine heath-like shrubs.

'The moss' *Rhacomitrium lanuginosum*

This is a most distinctive moss that in some localities can be a dominant component of the vegetation. It is a whitish or grey moss that forms **cushions or hummocks** varying 30–40 cm or more across, or is sometimes more extensive and forming contiguous colonies. Generally, the cushions are 10–12 cm deep. Its stems may be up to 10 cm long and they usually have numerous **short branches**. These branches are quite leafy. The colour of its **leaves** varies from a yellowish green to a brown but, from the outside, the whole cushion always has a very pale appearance because the tip of each leaf terminates with an extremely long, **silvery-white hair**. It is those hairs that are responsible for giving the plant its distinctive colouring. *Rhacomitrium lanuginosum* occurs in subalpine to low-alpine regions almost throughout New Zealand. It often forms extensive carpets or cushions on dry mountain slopes in the North Island, on the Volcanic Plateau, Mt Taranaki and southwards, sometimes also being found on lower hills. In the South Island it is found on the drier mountain ranges east of the main divide. It ranges from 700 to 1800 m. It often occurs on stony or rocky ground in open places, such as stable riverbeds. On Mt Taranaki it forms a most extensive and noticeable association in low herbfield and, locally, it is known simply as 'the moss', which appears to be the only known common name for it in New Zealand. It is a worldwide species usually being found in cold climates.

Tufted club-moss *Lycopodium australianum*

This is one of the few species of club-moss that inhabits subalpine areas. Unlike most other species of club-moss it is tufted and does not have creeping rhizomes or stems. Its size can vary according to habitat and it may be 5–20 cm tall. Under favourable conditions it may even grow to about 30 cm tall. Its rather rigid **stems** are erect and forked or they branch several times into stout and rigid **branches**. They are densely leafy with the narrow, **sterile leaves** being spirally arranged around the stem. Its leaves are about 1 cm long and they are sharply, almost pungently, pointed. The **sporangia**

are clustered in the axils of its **fertile leaves**, which are clustered near the tips of its stems. The tufted club-moss occurs in the North, South and Stewart islands and it also extends as far south as the subantarctic islands. It is not uncommon in alpine regions from the Raukumara Range and Rotorua southwards. Not infrequently, it grows in rocky places in snow tussock grasslands, herbfields, and subalpine scrub, especially in the higher rainfall regions. In the far south it descends to lower altitudes, particularly on Stewart Island. It ranges from 600 to 1700 m. This species also occurs in Australia and Borneo. The tufted club-moss is sometimes known as *Huperzia australiana*.

Alpine club-moss is a characteristic species of open heaths and grasslands, and is one of the commonest species, especially in the higher, hilly and mountainous regions. In impoverished and exposed habitats it may be no more than a centimetre or two high but in more sheltered conditions it will grow to about 40 cm tall. It has creeping underground **rhizomes** that are stout and branching, while its aerial **stems** are rigid and erect, their upper portions usually being densely branched. The **sterile leaves** are 3–5 mm long, they are spirally arranged and closely overlapping with their tips curving inwards. Depending upon their situation the leaves may be generally green but when growing in exposed and impoverished situations they may be yellowish or even bright orange. Their yellowish **strobili** are up to 7 cm long and erect on rather long stalks. They are either solitary or produced in groups of two or three. The size of the alpine club-moss can be quite variable depending upon habitat conditions. It occurs southwards from the Coromandel Peninsula, in the North Island, and occurs throughout most of the South Island and on Stewart Island. It usually grows in subalpine scrub, herbfields, grasslands and bogs. It is found from 200 to 2000 m. In the southern part of its range it descends to lowland districts. The alpine club-moss is easily recognised by its erect cones, incurved leaves and often by the orange colour that it assumes at higher altitudes.

Creeping club-moss *Lycopodium scariosum*

Among the native club-mosses this species is quite distinct and easily recognised on account of its flattened leaves and solitary strobili. It has rather stout, **creeping, branched main stems** that can be up to 2 m long, but are usually less. They are much-branched and their aerial stems may be up to about 50 cm tall but are often shorter. Its **sterile leaves** are up to 4 mm long and are flattened in the one plane. If they are closely examined, it will be seen that the leaves are of two sizes: the larger ones spread out from either side of the stem, while the smaller ones lie flattened, along the stems and between the larger leaves. The **strobili** are erect on

unbranched stalks and are about 2.5–5 cm long. In their young state they are a pale yellowish colour but with age turn brownish. The creeping club-moss occurs throughout the North, South, Stewart and Chatham islands. It also extends southwards to the subantarctic islands. While rare in Northland it is rather common throughout the rest of New Zealand, except for the dry east-coast areas of both main islands. It will often form thick, spreading colonies on banks as well as in scrub in montane and subalpine areas. Particularly in scrub country, it sometimes behaves as a semi-climber, scrambling for quite some height up through the stems of nearby shrubs.

Mountian kiokio *Blechnum montanum*

Mountain kiokio is one of five species that all have a rather similar appearance or characters. It is common at quite high altitudes in both main islands and also occurs on the Chatham Islands. It has rather thick **rhizomes** that are only shortly creeping, although in subalpine regions it may form conspicuous colonies. Its **stipites** are 8–20 cm long and are furnished with scattered, pale brown scales. The blades of the **sterile fronds** are 14–45 cm long by 6–25 cm wide; they are a deep green to bronze-green and have 6–20 pairs of oblong to slightly curved **pinnae** which are usually quite crowded with the longest ones being at or below the middle of the frond. The pinnae have tapering tips and their margins are finely toothed; the lowermost pinnae are scarcely shorter than the others. The **fertile fronds** are about the same length as the sterile ones and appear to be longer because of their tendency to stand erect. Their pinnae are also held in an upright position. Mountain kiokio occurs in the North Island from Mt Pirongia southwards, throughout much of the South Island and on the Chatham Islands. It usually grows in montane and subalpine regions to around 1100 m and in the south of the South Island may descend to sea level. Its habitat includes montane forests, subalpine scrub, grasslands and rocky areas. It also extends southwards to the subantarctic islands. At higher altitudes this species replaces *Blechnum novaezelandiae*, the common kiokio (not described).

Alpine hard fern *Blechnum penna-marina* **subspecies** *alpina*

One of the most ubiquitous of our native ferns and also one of the smallest species is the alpine hard fern. It occurs in many habitats in open country, riverbed shingles as well as alpine grasslands and scrublands. Especially at higher altitudes, it may be so dwarfed that it is often concealed among other vegetation. The alpine hard fern has **creeping rhizomes** that can spread quite far underground so that in suitable habitats it will form quite large colonies. It is easily recognised by its narrow fronds with the segments being attached to the midribs by their broad bases. The **stalks of the fronds** are also thin and wiry. The **fronds** are numerous and may be tufted at the tips of short branches or they may be spread along the whole rhizome. In exposed situations the fronds may be diminutive and flattened along the ground.

As is common with all members of the hard fern genus this species has two kinds of fronds: sterile and fertile. Both are long and narrow on wiry stalks. The **sterile fronds** are 3–25 cm long by 6–15 mm wide, are usually of a rather firm texture, dark green and simply pinnate. There are usually about 20–40 pairs of **pinnae** (often alternating) and they are mostly of equal length with those towards the base of the frond being of smaller size. The **fertile fronds** may be up to twice the length of the sterile fronds but their pinnae are more widely spaced and usually curved slightly upwards. The **sori** are numerous and cover the whole undersurface of each pinna. Alpine hard fern occurs throughout the three main islands, on the Chatham Islands, and on the subantarctic islands, extending as far south as Macquarie Island. It is common in lowland to high-alpine areas and may occur in open forest, scrub, open grasslands, moraines, herbfields and fellfields, ranging from sea level to 2000 m. In dry grasslands and at higher altitudes it is often quite dwarfed.

21

Bladder fern *Cystopteris tasmanica*

This is a delicate and rather unobtrusive fern that is often overlooked because it is frequently found growing in the clefts of rocks or tucked away among low scrubby places. It is summer-green and with the onset of winter completely dies down until the following spring. Bladder fern is a rather variable fern that has a semi-erect, short, creeping **rhizome** clad with pale, reddish brown scales. Its individual branches terminate with clusters of a few delicate fronds. In alpine habitats the **fronds** are frequently less than 15 cm tall, although in some situations they may be as long as 25 cm and are 1.5–7.5 cm wide. They are pale green or sometimes yellowish green, twice pinnate, with the primary **pinnae** being ovate to oblong; those on smaller fronds are divided into rounded lobes while those on larger fronds may be divided into secondary pinnae with blunt or rounded tips. The round **sori** are in one row along either side of the midrib and away from the margins of the pinnae. At first, the individual sori are enclosed in a rounded, bladder-like covering which disintegrates as the sori mature. It is this covering that gives rise to the fern's common name. The bladder fern occurs in montane to low-alpine areas in both the North and South islands, and also in Australia. In the North Island it is confined to mountains from the Kaimai Range to the Tararua Range including Mt Taranaki. In the South Island it occurs in mountain regions throughout and descends to lower levels in the far south. It ranges from 300 to 1200 m. It is probably more common than is realised, occurring in rock crevices, under rocky overhangs, sometimes in open grassy places or on rocks under beech forest. In the low-alpine zone it may also inhabit damp and shady overhangs of banks. While it occurs in humus-rich crevices in rocks, it can be more frequent and luxuriant on calcareous rocks and soils.

Bog pine *Halocarpus bidwillii*

As its common name suggests, bog pine often grows in peaty bogs, but it also grows in some quite dry locations. The 'Wilderness' area just near Lake Te Anau is a prime example. Usually, it is a quite **compact shrub**, of rounded habit, varying from 60 cm to about 3 m tall, and it can have a small **trunk** up to about 36 cm in diameter. In most situations its lower **branches** are spreading, although sometimes if they are low enough they may spread outwards and possibly root into the ground. The **leaves of young plants** are different from those of the adult, being 7–8 mm long, quite narrow and rather crowded, and

they spread outwards from the branchlets. The **leaves of the adult** are small and scale-like, densely overlapping and only about 1–2 mm long. Usually, they are dark green. The solitary male **strobili** are 2.5–3 mm long and are produced from the tips of the branchlets. The **female flowers** may be solitary or in pairs near the tips of the branchlets. Bog pine occurs in the North, South and Stewart islands in montane to subalpine scrub. It attains its northernmost limit on Mt Moehau at the northern tip of the Coromandel Peninsula and becomes more common as it progresses southwards. It altitudinal range is 600–1370 m but in western areas of both the South and Stewart islands it is found in lowland regions. Fairly high up on the Pisa Range in Central Otago, a lone, relict specimen of a rather large bog pine grows on a rock outcrop, indicating that in early times it may have once formed part of a forest on the now treeless Pisa Range.

Pygmy pine *Lepidothamnus laxifolius*

Pygmy pine is usually a prostrate shrub with slender, trailing branches, up to 1 m or more long, that will often form quite dense mats over rocks and among low-growing alpine vegetation. Sometimes it may also scramble up through other shrubby vegetation. On **juvenile plants** the small **leaves** are usually 5–8 mm long and they project out from the branchlets. Their colour varies from a very attractive glaucous colour to a bronzy green.

On **adult plants** the leaves become very small and scale-like, being reduced to about 1.5–3 mm in length, and they are closely flattened on the branchlets. The pygmy pine is often **unisexual** but, not infrequently, some plants will have flowers of both sexes on them. The small, up to 8 mm, **strobili** are produced on the male plants and will shed their pollen during early summer. The **female flowers** are produced from the tips of very short branchlets and develop into the fruits, typical of some podocarps, on which the blackish seed sits atop a **fleshy receptacle** that is generally red or orange. Although the fruits are said to ripen in the autumn, it is not uncommon to see some plants with ripe fruits on them during midsummer. The pygmy pine is found in the three main islands, occurring in the North Island from Mts Tongariro and Ruapehu southwards. In the South Island and on Stewart Island it is common throughout most mountain regions, apart from the drier eastern ranges. It usually occurs in subalpine to low-alpine areas where it grows in boggy or poorly drained sites. It can be common in sphagnum and cushion bogs and also grows in poorly drained snow tussock–herbfield. It generally ranges from 760 to 1500 m but on Stewart Island it descends to sea level. Some authors used to claim the pygmy pine as being the 'smallest known conifer', or the 'smallest conifer in the world'; however, *Microcachrys tetragona*, of Tasmania, is probably equally deserving of that title.

Mountain totara *Podocarpus nivalis*

Mountain totara is a rather variable species that, in some localities, may be a quite prostrate shrub while in others will form a dense, bushy plant up to some 2 m or so tall. While it is generally seen as a component of subalpine scrub and similar habitats, it can also occur as an understorey plant in the higher altitude alpine forests. Its **leaves** are closely set around the branchlets and are more or less spirally arranged. They are 8–16 mm long by 1.5–3 mm wide, bronze to deep green or yellowish to deep green and the leaf tips have small, pungent points. Individual plants are **unisexual**; the **strobili** of male plants are 7–25 mm long and they may be solitary or in groups of up to four. Normally, they are quite obvious and are produced during early summer. On female plants the **flowers** are solitary and also

produced from the axils. From mid to late summer the green, nut-like **seeds** sit on the tops of succulent, swollen receptacles. These receptacles are actually the foot-stalks of the seeds, and their colour varies from orange to red. Mountain totara commonly occurs in subalpine scrub, mixed snow tussock–scrub and among the shrubby vegetation that colonises old moraines. It will also occur in higher altitude alpine forests. The species occurs mainly in subalpine to low-alpine regions throughout the North and South islands, from Mt Moehau at the northern tip of the Coromandel Peninsula, southwards to Foveaux Strait. In the North Island its distribution is inclined to be local, but it occurs throughout most of the South Island mountain regions. It ranges from 700 to 1500 m. Mountain totara is also referred to as snow totara, which is nothing more than a translation of its scientific specific name.

RANUNCULACEAE

Of the native species of buttercup this is one of the smaller species but nonetheless a very attractive one that is quite widely distributed throughout the mountain regions of the South Island. It is a smallish, slender species, usually 5–15 cm tall when in flower. Its rootstock is short and stout and not visible above ground. It has a distinctive rosette of **leaves** at the top of its rootstock, most being 2–12 cm long. The individual leaves may be quite variable with two to six pairs of leaflets varying from three- to five-lobed, or irregularly and deeply cut into narrow, pointed segments. Its slender **flower stems** vary

from one to several in number and may be produced in succession over the flowering season; they are usually longer than the leaves. The single **flowers** are 1–2 cm in diameter and are typically buttercup yellow. Flowering occurs during November and December. *Ranunculus gracilipes* occurs in subalpine to high-alpine areas of the South and Stewart islands. It is fairly widespread from about south of the Nelson Lakes National Park and the Awatere Valley to western Southland and eastern Fiordland, and then to Mt Anglem on Stewart Island. It ranges from 700 to 1800 m but in the far south descends almost to sea level. Particularly in some areas, it forms quite dense colonies and is then a delight to behold.

This remarkable species of buttercup occurs only on the vast mobile shingle screes that characterise the drier eastern ranges of the Southern Alps. It is a summer-green herb that dies down with the onset of winter and does not emerge again until the spring thaw, when shingle screes are free of snow, and it is well able to survive in one of the harshest of alpine environments. *Ranunculus haastii* arises from a thick, fleshy **rhizome** that is buried deep in the scree where its roots are perpetually moist. Its fleshy **leaves** are glaucous-grey and so perfectly do they blend with the grey stones of the shingle scree that the scree appears to be devoid of plant life. It is only when the plant flowers and produces its gorgeous golden goblets that it really becomes noticeable. The leaves are deeply cut to their bases into about five or seven lobes, which are again further divided. They vary in height, 5–15 cm, and the blade of the leaf is 5–10 cm in diameter. The **flowering stem** is thick and fleshy and, frequently, there is only one flower per stem, but on vigorous plants there may be up to six flowers. The bright yellow **flowers** are about 4 cm in diameter and usually have 5–20 waxy petals. Flowering occurs during November to January. After flowering its globose **fruiting heads** are quite conspicuous. *Ranunculus haastii* is found in low- to high-alpine areas of the central and eastern ranges of the Southern Alps from the Seaward Kaikoura Range, in Marlborough, to the Takitimu Range in Southland. It is restricted to moving and semi-stable shingle screes, and debris slopes, mainly in greywacke areas, and can be found from 1000 to 2000 m. Plants are sometimes ravaged by hares and other browsing mammals. Subspecies *piliferus* differs by having more rounded segments to its leaves and long silky hairs on their upper surfaces. It occurs only on the Eyre Mountains and some adjacent ranges (see lighter shading on map).

Korikori *Ranunculus insignis*

This beautiful species, formerly known as *Ranunculus monroi*, is without doubt the finest of the yellow-flowered native buttercups and it occurs in alpine regions of both the North and South islands. Korikori is a complex species that includes some quite small forms and some quite large ones. The rather diminutive forms may be no more than 10 cm tall, whereas the more robust plants may be up to 60 cm or even 90 cm tall. It has **tufted** growth and in its larger forms the thick, evergreen basal **leaves** are on long stalks 10–25 cm long. The leaf blade is 15–22 cm in diameter, its upper surface deep green and shiny; it is heart-shaped or kidney-shaped and is conspicuously fringed around the margin with brown hairs.

The leaf margin is also coarsely toothed. The branched **flower stem** usually carries numerous golden-yellow **flowers** that may be up to 5 cm in diameter. They usually have 5–10 petals. On the smaller forms there may be just one flower per stem. Flowering occurs between November and February. Korikori inhabits shady areas of grassland, herbfield, subalpine shrubland, sheltered bluffs and rock outcrops. It occurs in subalpine to low-alpine areas in the North Island from Mt Hikurangi to the central volcanoes then southwards to the Ruahine and Tararua ranges. In the South Island it is most common in the mountains of Nelson and Marlborough with the smaller forms occurring from North Canterbury to about the Two Thumb Range in South Canterbury. It ranges from 700 to 1800 m.

Mountain lily *Ranunculus lyallii*

The mountain lily is one of the most beautiful plants in the alpine flora of the South Island. It is the largest of the New Zealand species of *Ranunculus* and is justly acclaimed as being one of the most magnificent buttercups in the world. In sheltered sites it may stand up to 1 m or more tall. When not in flower the mountain lily is easily recognised by its bold saucer-shaped, leathery **leaves** of a deep, shiny green. Sometimes, after a mountain shower, they may be seen partially filled with water. The leaves are 10–40 cm in diameter and are on stalks up to 30 cm long. When the mountain lily flowers it can be a breathtaking sight. Its branched **flower stems** carry numerous flowers of the purest white. Each **flower** is 5–8 cm in diameter and has 10–25 overlapping petals. The centre of the flower, where the seeds are formed, is greenish and is surrounded by a circle of golden-anthered stamens. After flowering the green seed-heads soon ripen and the seeds fall. At lower altitudes flowering usually commences about late November and, as altitude increases, it becomes progressively later until the last of the flowers persist until January. Mountain lily occurs in the South and Stewart islands, in subalpine to low-alpine areas in the higher rainfall regions, about and west of the main divide, from Mt Buckland near Westport southwards.

It is more plentiful to the south of Arthur's Pass but is rare on Mt Anglem on Stewart Island. It is usually found in snow tussock–herbfield, alongside streams, in wet hollows and flushes and on rock bluffs and faces. Feral deer and chamois are enemies of the mountain lily. *Ranunculus lyallii* is also known as the Mount Cook lily and the mountain buttercup. The latter name arose because purists insisted that it is correctly a buttercup and not a lily.

RANUNCULACEAE

Formerly known as *Caltha*, there are two species of *Psychrophila*: *P. obtusa* with white flowers and *P. novae-zelandiae* with yellow flowers. The former is confined to the South Island while the latter occurs in the North, South and Stewart islands. *Psychrophila obtusa* is a low-growing plant with short, creeping, whitish **rhizomes**. It is rarely more than about 5 cm tall and, although it will form bright green colonies up to 50 cm across, when not in flower it is not always easily distinguished. Its deep green **leaves** are seldom more than 1 cm in diameter, prominently lobed at their bases with the lobes being almost as long as the main part of the leaf blade. Their margins are

distinctly notched with rounded teeth. Its **flower stem** is very short but elongates after flowering. It has only a single white **flower** which is about 12 mm in diameter and has five more or less obovate sepals with blunt or slightly pointed tips. It can be one of the first alpine plants to come into flower during spring, and it can sometimes be seen coming into flower almost under the receding edge of melting snow. Depending upon the season and altitude, flowering is usually between November and January. *Psychrophila obtusa* occurs in low- to high-alpine areas of the South Island from Canterbury to Central and western Otago and eastern Fiordland. It ranges from 1200 to 1800 m. It is common in wet hollows or on snow-banks, alongside streams and other permanently wet sites.

Penwiper *Notothlaspi rosulatum*

This singular and remarkable plant is mainly confined to the shingle screes that characterise the drier eastern mountains of the South Island. It is a **monocarpic** plant, which means that it grows one to two or more years before flowering. It then flowers, only once, during its second to third (maybe even not until its fourth) year and then seeds before it finally dies. The penwiper is so named because it resembles an old-fashioned, Victorian cloth pen-wiper that was formerly used to wipe the ink from the nib pens used in those days. It is a deeply **taprooted** plant, the leaves of which form but a single rosette that may be up to about 8 cm in diameter. Its numerous, shovel-shaped **leaves** overlapping like the tiles of a roof closely match the grey colour of the surrounding stones. They are thick and fleshy, slightly ribbed and slightly hairy. Initially, the leaves grow flat, but once the plant reaches the flowering stage they become slightly convex so that the rosette forms more or less a dome or umbrella shape with only their outer edges touching the stones of the scree. On larger plants the stout **flowering stem** may be up to 25 cm long and the inflorescence is cone-shaped with numerous flowers. Its **flowers** are creamy-white, about 1 cm in diameter and deliciously scented. Flowering plants may usually be found between November and January. After flowering the flattened **seed capsules** develop and, at maturity, they may be up to 2.5 cm long. They are shaped like an inverted heart. The penwiper occurs in subalpine to low-alpine areas of the drier greywacke mountains, east of the main divide in the South Island, from Marlborough and eastern Nelson to North Otago, ranging from 800 to 1800 m. It usually grows on the finer and more stable screes, or frost-eroded areas that are predominantly stony.

Mountain violet *Viola cunninghamii*

This is an attractive little species of violet that may be recognised by its **tufted habit of growth**, with usually either one or several tufts. A second, similar species (*Viola lyallii*) usually occurs in damp areas and differs from *V. cunninghamii* by having a creeping habit of growth and the bases of its leaves being conspicuously heart-shaped. The bases of leaves of *V. cunninghamii* are generally more or less deltoid, and either somewhat square-cut across to slightly tapering to the stalk. *Viola cunninghamii* is a summer-green herb and its glabrous **leaves** are more or less triangular in shape with blunt tips. Their **petioles** are quite slender and, depending where it is growing, they may be quite long. The **flowers** of *V. cunninghamii* are like those of a typical violet, about 1–2.5 cm in diameter, and they typically stand above the leaves on slender stalks that may be 5–10 cm long. They are white with a number of purple lines in the throat and on the lowermost petal. There is also a touch of yellow or greenish yellow in the throat. Unlike some exotic and garden-hybrid violets, the native violets do not have scented flowers. The **seed capsule** is about 7–10 mm long and is green when ripe. At that stage it splits open to distribute its seeds. Flowering may occur any time between November and March. In later flowering, following early summer, violets often produce what are known as **'cleistogamic' flowers**; that is, flowers that never open but still have the ability to pollinate themselves and produce capsules containing viable seeds. *Viola cunninghamii* occurs from lowland to high-alpine regions in the North, South and Stewart islands. It is found almost throughout the whole range, from sea level to 1800 m. In alpine areas it occurs in unstable rocky sites, in damp or moist places in grassland, stabilised riverbed, shrubland, and rocky places.

Sundew *Drosera arcturi*

Sundews are among New Zealand's few carnivorous plants and they usually grow in boggy habitats, where there is a nitrogen deficiency, hence their carnivorous habit. *Drosera arcturi* is a summer-green herb that is easily recognised because of its long, strap-shaped leaves and single flowers that stand clear of the leaves on relatively long stalks. Being carnivorous, sundews have a diet of insects which become entangled in, and stuck to, the long, sticky hairs that characterise the leaves of these plants. The **leaves** are up to 5 cm long by about 8 mm wide and the **glandular hairs** with which they are clad are of varying lengths so that small flies and other insects

may much more readily become entangled in them. Each hair secretes a droplet of clear, **sticky fluid** from its tip. Once an insect becomes caught by the fluid, the hairs along the margins gradually curl over the insect and then glands on the leaf pour out a special fluid that may be likened to a kind of **gastric juice**. After a few days the insect will have been completely absorbed, apart from any indigestible portions. In this way the sundew is able to obtain the essential nitrogen that may be lacking in the soil in which it grows. Its usually solitary white **flowers** may be carried on stems up to 15 cm in length although, generally, they are shorter. Each flower is 10–12 mm in diameter. Flowering occurs between November and March. The **seed capsules** are black and about 1 cm long. *Drosera arcturi* is found in peat bogs, around the margins of tarns and on wet clay banks in montane to low-alpine areas, ranging from 300 to 1500 m. It occurs in the North, South and Stewart islands, being widespread from the central North Island southwards. It is also found in Tasmania.

Sundew *Drosera spathulata*

This is a smaller species of sundew than *Drosera arcturi* (see previous entry) and in some localities the two species may be seen growing together. *Drosera spathulata* may be recognised by its rosette of **leaves** generally growing flat on the ground and spread out in circular fashion. Their **petioles** are up to 10 mm long and they widen to the rounded leaf blade which is usually about 5 mm or so in width. Both the petiole and the blade are generally quite reddish while the blade is covered with stalked, glandular hairs that are tipped with the glistening, sticky globules that help to trap small insects. Plants growing at higher altitudes are distinctly smaller than those growing at lower levels. There may be just one or a couple of **flowering stems**, each bearing just one to two flowers or, at lower altitudes, several flowers. Its white **flowers** are about 6–8 mm in diameter. Flowering is usually between November and February. After flowering small black **seed capsules** are formed. *Drosera spathulata* occurs in the North, South and Stewart islands where it is usually found from lowland to low-alpine areas. It usually inhabits peaty ground, subalpine bogs and thin clay soils, often with *Drosera arcturi* or other sundews. Although rather widely spread throughout the country, this species can be more local in its occurrence. It ascends to 1400 m.

This is an alpine cushion plant that is quite
thickly tufted and, unbelievably, it actually
belongs to the same family as the chickweed
(*Stellaria media*) and carnation (*Dianthus*).
It arises from a **central taproot** and forms
small, dense **cushions**. Generally, they may
be 10–15 cm in diameter. Each cushion is
made up of tightly tufted stems that are
densely covered with overlapping leaves.
These **leaves** are 5–7 mm long, awl-shaped
and gradually taper to long, stiff and sharp

points. Inside the cushions the lowest portions of the stems are
clad with the old leaves, which are often still very much intact. The
flowers, 7 mm long, are shorter than the uppermost leaves and

are usually buried amid their needle points. The flowers have five
green sepals and do not have any petals. Flowering occurs during
November and December. *Colobanthus acicularis* occurs in both the
North and South islands, usually on dry, rocky places especially
in crevices on bluffs and rock outcrops. In the North Island it has
been recorded in the Ruahine Range. In the South Island it is found
on the drier mountains east of the main divide from Nelson to at
least Mid Canterbury and possibly in the mountains further south.
It ranges from 500 to 1800 m.

CARYOPHYLLACEAE

This is another of the distinctive species that inhabit the shingle screes that occur on the mountains running along the eastern side of the main divide of the Southern Alps. It is a most unusual plant; the aerial parts of which are grey, similar to the coloration of the stones of the screes which it inhabits. Like all scree plants its **roots** go deeply into the moist strata deeper within the scree. The **stems** that ascend to the surface are extremely delicate and easily broken yet, strangely, they are able to withstand the continual slippage of the surface layer of stones without any ill effects. As the stones gradually move downhill, thus causing the above-ground parts of the plant to move down as well, its stems simply lengthen sufficiently to

enable its top parts to more or less stay where they were. *Stellaria roughii* forms clumps of growth that may be up to 20 cm across and up to 5–10 cm high. Its narrow, pointed **leaves** are in pairs on much-branched stems and they may be up to 2 cm long by no more than 3 mm wide. As well as being grey, the leaves are soft and rather fleshy. Its solitary, white **flowers** are produced from the ends of the branches, and they are surrounded by five leaf-like **sepals**. Flowering may occur from December to February. As is usual with scree plants, *Stellaria roughii* is usually scattered quite sparsely over the surface of the scree and the plants are not easily discerned. The species was named after a Captain Rough who discovered it in the mountains of Nelson. It is found in low- to high-alpine areas of the South Island from Nelson to western Southland, and ranges from 1000 to 2000 m. It is confined to shingle screes and similar loose, stony sites.

GERANIACEAE

This is one of the commonest of the native geranium species. It normally forms a clump that arises from a single, stout rootstock that may branch to form more than one crown. It usually exists in two colour forms: one has green foliage and the other has bronze to purplish-bronze foliage, both belonging to the same variety. In many localities bronze-leaved plants appear to be the commonest colour form and can be seen in company with the green-leaved form. *Geranium sessiliflorum* usually has a long, woody **taproot** that descends deeply into the stony soil of its habitat. The **foliage growth** is usually quite tightly clustered, in a rosette, around the crown of the plant. Its **leaves** are on quite long, slender petioles and the leaf blade is usually 1–2 cm in diameter. The leaves are kidney-shaped, with five- to seven-lobed margins, and there are often a few hairs on the surface of the blade. Its **flowers** are white (occasionally pinkish) and are produced singly on very short stalks, but sometimes they may be up to 3 cm long. The specific name of *sessiliflorum*, which means 'stalkless', is actually a bit of a misnomer because, although the flowers may appear to be without stalks, they are actually on very short stalks that elongate as they age. Each flower may be up to 1 cm in diameter. *Geranium sessiliflorum* occurs throughout the North, South and Stewart islands from sea level to high-alpine areas, ranging up to 1700 m. It is more common in lowland to subalpine grasslands and coastal areas than it is in the alpine zone. In alpine areas it frequently occurs in rather dry, open, stony sites, particularly in snow tussock grassland, induced herbfield (created by burning, grazing or other means), semi-stable scree and old riverbed country.

Muehlenbeckia axillaris

Muehlenbeckia axillaris is a **creeping shrub** that, depending upon where it is growing, forms either quite large mats up to a metre or more across, or smaller, more straggly patches. Because of its creeping habit its dark, wiry **stems** may be above ground or, more frequently, they creep underground and thus help to stabilise the stones of streambeds and open areas. Its small, dark green **leaves** are of different shapes but, generally, they are rounded and have short or long, slender stalks. Their undersurfaces are paler. The leaf blade is usually 3–5 mm in diameter. There are separate male and female **flowers** on the same plant and they occur in the leaf axils, either singly or in pairs. Each flower is about 4 mm in diameter and is a yellowy green in colour. Flowering may occur between November and April. While the flowers are noticeable, it is the **fruits** that are the most prominent feature of this plant. After pollination, the tepals of the flowers enlarge to become succulent and glassy so that they are quite translucent. They are united at their bases to form more or less of a cup in each of which sits a shiny black seed. A well-fruited plant is a fine sight. Sometimes the seeds are formed without the development of the succulent tepals. Fruiting generally continues throughout the length of the flowering season. *Muehlenbeckia axillaris* is rather widespread in the mountains of the North Island from eastern and central areas southwards. In the South Island it occurs throughout most mountain regions. It is usually found in montane to low-alpine areas and is common in damp or dry stony sites, particularly on gravel riverbeds and streambeds and on moraine, and it also occurs in open grassland and herbfield. It ranges from 300 to 1500 m.

Willow-herb *Epilobium crassum*

The genus *Epilobium* is quite large and contains at least 37 native species. About 18 of them inhabit alpine areas, but not many are likely to be noticed by trampers and others who visit those areas. *Epilobium crassum* is a distinct and attractive species of willow-herb that is very easily recognised. It has compact growth with shortly creeping, **stout stems** that are 5–15 cm long and root into the ground. Its **leaves** are closely placed or crowded along the stems, but are wide-spreading. They are 3–3.5 cm long by up to 1.4 cm wide, quite thick and fleshy; their upper surfaces are a bright, shiny green while their undersurfaces are reddish purple. The tips of the leaves are blunt and there is a suggestion of shallow toothing around their margins. The **flowers** vary from white to pink and are about

8–10 mm in diameter, with the petals being split to about one third of their length. Flowering takes place during the summer months from about December to February. Its long, stout **capsules** are held at the ends of stiff stalks that are about as long as the capsules. *Epilobium crassum* occurs in the South Island on the drier greywacke ranges, east of the main divide, from the St Arnaud Range of Nelson and the Molesworth in Marlborough to North Otago and Lake Wanaka. It is found in low- to high-alpine areas where it grows on fine, relatively stable shingle screes, scree margins and broken rocky places in fellfields. It ranges from 1000 to 1800 m.

ONAGRACEAE

Epilobium macropus has rather slender growth and is often quite obvious because of its habitat in damp or wet situations, often forming quite large floating masses in slowly flowing mountain streams. Its dark purple, slender **stems** may be up to 25 cm long, and they root into the damp soil in which they grow or into the stream water in which they float. They usually become quite erect towards their tips. The **leaves** are ovoid, usually about 1–1.5 cm long by about 4–10 mm wide. Their margins are obscurely or distinctly toothed. It has easily noticed white **flowers** along its stems. The flowers are 9–12 mm in diameter and are usually produced over quite a long period from November to April. After flowering, dark-coloured **seed**

capsules develop, usually about 4–5 cm long. *Epilobium macropus* occurs in the North Island from about Mt Tongariro southwards, and throughout many of the mountain areas of the South Island, extending as far as Fiordland. It ranges from 600 to 1500 m and grows in montane to low-alpine regions.

Willow-herb *Epilobium pycnostachyum*

This species of willow-herb is quite distinct from the other members of the genus because of its **tufted habit of growth** and the fact that it is usually found only on shingle screes and similar areas. At its base it is semi-woody and during the winter it tends to die down to its tufted rootstock, re-emerging in the spring once the snow has thawed from the scree. It has an extensive, much-branched **root system** that penetrates quite deeply into the scree. The numerous **stems** are reddish to purplish or green; unbranched, semi-erect to erect; and tend to be intertwined, especially near where they emerge from its base. Its light- to medium-green (sometimes reddish) **leaves** are close-set along the stems and spreading to somewhat ascending. They are about 1–2 cm long by 2–4 mm across and are coarsely toothed around their margins. The numerous white **flowers** are 8–9 mm in diameter and are produced from the leaf axils along the upper part of the stem. Flowering is usually during December and January. The red **capsules** are 1.2–2 cm long. *Epilobium pycnostachyum* occurs in low- to high-alpine regions in both the North and South islands from the Kaweka Range and Mt Ruapehu southwards. In the south it attains its southernmost limit on Mt Ida in Otago and on the Thomson Mountains in Southland. It ranges from 800 to 1800 m. It is often fairly widespread especially on the South Island's dry eastern greywacke mountains, but it does extend as far westward as the main divide. It may be common as an inhabitant of relatively mobile screes but it can also occur on more stable, eroded sites in fellfield.

THYMELAEACEAE

Formerly included in the genus *Drapetes*, this plant and its close relatives were more recently classified as members of the genus *Kelleria*, named after a Engelhardt Keller, author of an 1838 book on wine. Of the 11 species of *Kelleria*, nine are considered to be alpine plants. They are all low-growing, trailing or cushion subshrubs or shrublets. The species differ in their habit of growth, the degree of hairiness on their stems and whether their stems are glaucous or not. Of the various species, *Kelleria dieffenbachii* is the most widespread and the one most likely to be encountered. It often forms quite **dense mats** up to 30 cm or more across but in some habitats its growth may

be looser. Its prostrate **branchlets** are 7–30 cm long, clad with small leaves after the fashion of a whipcord hebe and they often root into the ground. The grey-green **leaves**, 2.5–3.5 mm long by 0.5–1 mm wide, are tightly flattened to the branchlet so that they closely overlap. They have blunt tips and their margins are fringed with small white hairs. The small **flowers** of *Kelleria* are similar to those of *Pimelea*, to which it is related. Usually, the flowers are produced from the tips of the branchlets in heads of three to eight flowers. The individual flowers are quite small, being no more than 3–4 mm in diameter. They are either perfect (male and female parts in the one flower) or are unisexual, all being on the one plant. The small, ovoid **fruits** are fleshy and coloured orange. Flowering and fruiting take place during December and March. *Kelleria dieffenbachii* occurs in subalpine to low-alpine areas throughout the North, South and Stewart islands from Mt Moehau on the Coromandel Peninsula southwards. It grows in open shrublands, tussock grasslands and fellfields. It ranges from 600 to 1500 m.

Kelleria laxa

This species is fairly easily distinguished from *Kelleria dieffenbachii* (see previous entry) by its loose habit of growth and its rather pale green leaves, which are 4–7 mm long and are wide-spreading around the branchlet. Generally, it forms **loose patches** rather than the tighter, compact mats of *Kelleria dieffenbachii*, and may often be no more than 20 or 30 cm across. The branchlets are prostrate and root into the ground. Its spreading **leaves** are usually a

light or medium green, flat, quite narrow and about 4–7 mm long. Seen through a hand lens their margins are fringed with hairs. The whitish or rather creamy **flowers** are quite small, no more than

about 2 mm long, and there are three to eight per head at the tips of the branchlets. Flowering is usually between January and February. *Kelleria laxa* occurs in both the North and South islands from the Volcanic Plateau, the Kaimanawa Mountains and Ruahine Range, southwards to Nelson and northern Westland and thence to the Blue Mountains of Otago and the Longwood Range in Southland. It occurs in subalpine to high-alpine regions in grassland and fellfield, ranging from 900 to 1700 m.

CORIACEAE

The mountain tutu is a summer-green, semi-shrubby plant that grows up to 50 cm or so tall and often forms quite large patches or colonies. As with its larger relations, all parts of the mountain tutu are quite poisonous. Its slender or stout, branching **rhizomes** spread underground and from them arise slender **branches** and **branchlets** with the latter being ascending. They bear numerous opposite, or almost opposite, small **leaves** that give the branchlets a very feathery appearance. The leaves are 7–10 mm long by 1–2 mm wide, are deep green and are quite shiny. Mountain tutu's small greenish **flow-**

ers are produced on racemes 3–5 cm long and are quite widely spaced. As the **seeds** form, the flower petals turn black and enlarge to become quite succulent to enclose the seeds so that they appear to be berry-like. The plant flowers and fruits between November and February. The top growth usually dies down about May and new growth commences to reappear in September. Mountain tutu grows in montane to sub-alpine areas in the South Island from about north-western Nelson southwards and also on Stewart Island. It is common in stony areas, on moraines, along stream-banks, in scrub and tussock grassland, mostly about and west of the main divide. The species ranges from 500 to 1500 m.

Biddy biddy *Acaena glabra*

This is one of the most distinctive of the *Acaena* species because, unlike the others, it does not have an aggressive, wide-spreading habit and also the seed-heads do not have barbed spines. It is also almost completely without hairs on any part. These characters make it easy to recognise; in particular, the large, spineless seed-heads at once distinguish it. It forms **patches** up to about 30 cm or so in diameter, and of rather a loose nature, with the tips of the branchlets becoming slightly erect. Its **leaves** are up to 5 cm long and have 6–10 pairs of leaflets with one slightly larger terminal leaflet. The upper surfaces are brownish to yellowish green and shiny, while the undersurfaces are more or less glaucous. The **flower heads** are quite large, 1.5–2 cm in diameter, green to purplish, with white stamens, and the heads are produced on stalks up to 10 cm tall. Each **seed** on the head actually has four rudimentary spines (without barbs) that are seldom more than 2 mm long. Flowering usually occurs between October and December and the seed-heads mature between December and February. *Acaena glabra* is confined to the South Island in subalpine to low-alpine regions where it is sometimes quite widespread on the greywacke mountains along and east of the main divide. It ranges from 600 to 1400 m. The species is commonly found on scree margins, stable screes, in open places in eroding grasslands and on riverbeds. All *Acaena* species are commonly referred to as biddy biddy (or similar variations), which is a corruption of their Maori name of piripiri.

45

Biddy biddy *Acaena inermis*

Acaena inermis is one of the smaller species of biddy biddy and, depending upon the part of the country, it is also a somewhat variable species. It is a creeping and rooting **mat-forming** plant that can often extend over quite large areas. Its **stems** may be slender or stout, occasionally subterranean, and its branchlets can be almost concealed among the surrounding vegetation or somewhat ascending and up to 4 cm long. Particularly when growing in dry grassland, it can be quite diminutive but, in better conditions, it is larger and more noticeable. Its **leaves** are usually greyish or brownish grey, around 1.5–1.7 cm long and have 11–15 leaflets. The terminal leaflet is usually larger than the others. The margins of the leaflets have 7–12 teeth around them. When they first appear the **flower heads** are

4–6 mm in diameter and they are produced on stalks that are usually 1–5 cm long. After pollination they enlarge to about 2.5 cm, including the spines, if present. The spines of this species do not have barbs, which means that they are not able to attach themselves to wool and other fibres. Especially along the eastern side of the South Island, *Acaena inermis* quite often has flower heads without any spines at all, but in some South Island localities, plants with bright crimson, barbless spines may also be seen growing near typical plants that have no spines whatsoever. The specific name of *inermis* means unarmed and probably refers to the fact that its spines have no barbs. Flowering occurs between November to January while fruiting is normally between January and April. *Acaena inermis* occurs in montane grasslands of the Kaimanawa Mountains and other ranges of the central North Island. In the South Island it is common in the drier eastern areas from Marlborough southwards, where it may be found in grasslands and riverbeds and mainly along the eastern side, and in herbfields up to about 750 m. All *Acaena* species are commonly referred to as biddy biddy (or similar variations), which is a corruption of their Maori name of piripiri.

Geum cockaynei

This native geum, formerly known as *Geum parviflorum*, is a tufted, **rosette-forming** herb and is the commonest of the native species. It arises from a stout, woody **rootstock** that may be branched and have several crowns. It is easily recognised by its wide-spreading, bright to deep green **leaves** that have shiny upper surfaces. They are 6–15 cm long and always have a very large terminal leaflet along with up to 15 pairs (usually fewer) of quite small leaflets along the midrib. The terminal leaflet has distinctly toothed margins and it is hairy on both surfaces. Its hairy **flower stem** may be branched and up to 30 cm tall. The **flowers** are white and up to 1.5 cm in

diameter. They usually appear during December and January. *Geum cockaynei* is found in both the North and South islands from about Mt Hikurangi and the Ruahine Range southwards. It occurs in subalpine to low-alpine areas and generally inhabits moist sites in snow tussock grassland and herbfields, shady rock faces and bluffs, and on shady ledges. It ranges from 800 to 1700 m. While it will grow in a variety of situations, it is most noticeable when it grows on bluffs and rock faces. It is absent from the central Volcanic Plateau and Mt Taranaki, as well as Central Otago.

Carmichaelia monroi is a small shrub that is usually found in the drier regions east of the South Island's main divide. It is a dwarf, much-branched **shrub** that grows from a strong single taproot. Usually, it forms hard, flattish plants up to about 20 cm in diameter or slightly rounded hummocks up to about 15 cm high. Its hard and rigid **branchlets** are 2–4 mm wide, quite flattened and grooved. Their tips are blunt to subacute, but as they are often browsed by hares and rabbits, they may have a cut-off appearance. Consequently, the plant often does not exhibit its true image. Adult plants are completely leafless, the stems carrying out the functions of leaves. The **flowers** are pinkish purple or reddish purple and produced in clusters of

two to five from small notches along the sides of the branchlets. They are usually produced during November or December. The flowers are followed by small **seed-pods** about 1.5 cm long that are very dark brown to almost black. The species is confined to montane to low-alpine regions of the South Island from Marlborough to North Otago, ranging from 800 to 1500 m. It occurs in the drier mountain areas east of the main divide, usually in stony or rocky and well-drained open habitats in open or depleted low tussock grassland.

Golden Spaniard *Aciphylla aurea*

Golden Spaniard is a handsome species that cannot fail to be noticed. It is a large herb that forms clumps of single to several **rosettes** or tufts, with each clump up to a metre or more in diameter. Each rosette is made up of crowded, rigid, spine-tipped **leaves**. The tips of these leaves point outwards in all directions and seemingly present a formidable armament against browsing predators. However, these plants are very palatable to hares, rabbits and other browsing mammals, and they have no difficulty in penetrating the plants' defences to feed on them. In spite of that, some people have even postulated that the spines were a defence against the browsing of the now extinct moa, although the most likely explanation for the origin of the spinescent leaves is that they evolved as a protection against wind, drought and excessive exposure to strong sunlight, as is evidenced by plants of like form growing under similar conditions in other countries. The **leaf blade** is twice divided so that it is narrowly fan-shaped in outline and is more or less yellowish green, or sometimes golden, with the midribs usually being greenish to pale yellow. Its **flowering stems** are up to a metre or so tall and, apart from their golden colour, they are notable because the small clusters of yellowish **flowers** are protected by long, narrow, intermeshing spines that also protect them against the effects of desiccating winds. Individual plants are either male or female. Flowering is usually during December and January. Golden Spaniard occurs in montane to low-alpine areas of the South Island from Nelson and Marlborough to northern Southland, ranging from 300 to 1500 m. It is found in drier areas mainly east of the main divide in mixed snow tussock–scrub, tussock grassland and among rocky outcrops. It is also known as taramea and formerly as bayonet grass, speargrass or wild Spaniard.

49

Speargrass *Aciphylla colensoi*

APIACEAE

This species is somewhat similar to *Aciphylla aurea* (see previous entry) but differs mainly in the midribs of the leaf divisions being very prominent and distinctly reddish to orange or golden-yellow. It can form quite large **clumps**, up to 90 cm in diameter and 40–50 cm in height. As with *Aciphylla aurea*, it has a long and stout **taproot** that penetrates very deeply into the ground and it usually produces several **rosettes** of leaves.

Its rigid **leaves** are usually green or greyish green and their margins are distinctly serrated when examined through a hand lens. The **flowers** are produced on long stems, up to a metre or so tall, and they are usually conspicuous because of their orange colour. The flowers are quite small and are in dense clusters, the male and female flowers being on separate plants. The male plants are more showy. Long narrow spines project out from among the flowers and present quite a formidable appearance. Flowering is usually during November and December. The flowers are pollinated by small beetles and weevils. *Aciphylla aurea* occurs in the North and South islands from Mt Hikurangi to about Mid Canterbury. It is often widespread in subalpine to low-alpine areas and may grow in subalpine scrub, mixed snow tussock–scrub, herbfields and grasslands. Generally, it prefers to grow in slightly moister situations. It ranges from 900 to 1500 m. The plant's Maori name of taramea may also refer to any of the similar large species of *Aciphylla*. As with the other large species it has a number of common names with speargrass and Spaniard probably being the most commonly used.

Speargrass *Aciphylla ferox*

This is somewhat similar to the two previous species (*Aciphylla aurea* and *A. colensoi*) but is generally a little smaller in its growth. It forms single or multiple **rosettes** and individual plants may be up to about 60–75 cm in diameter by about 40 cm tall (excluding the flowering stems). Its **leaves** are characteristically curved, which is usually a good point of recognition. They are green to olive-green and the segments are rather stout and relatively broad compared with those of *A. aurea* and *A. colensoi*. A further point of identification is the long stalk between the sheathing base of the leaf and the leaf blade. This stalk is distinctly longer than those of the aforementioned species.

The **flowers** are in clusters on the flowering stem, which is at an inclined angle as it projects outwards, as well as being slightly curved, rather than erect as in other large species. The flowering stem is about 75 cm long. The flower clusters are more or less hidden by its quite large, flattened spines that are greenish to orange and all pointing upwards so that the whole flowering stem has quite a narrow aspect. Flowering is generally between December and January. *Aciphylla ferox* is somewhat similar to *A. horrida* (not described in this book), the range of which occurs to the south of *A. ferox*, and extending as far south as Otago. *Aciphylla ferox* occurs in Nelson and Marlborough where it grows in subalpine to low-alpine areas, ranging from 600 to 1400 m. It inhabits short subalpine scrub, mixed snow tussock–scrub and is sometimes conspicuous in short snow tussock–herbfield.

Aciphylla monroi and *A. similis*

Aciphylla monroi is a tufted herb that is one of the smaller members of the speargrass genus. It usually has several or many **rosettes** arising from its **main taproot** and is up to about 10 cm tall. It can have relatively few leaves or they can be more numerous and crowded on each rosette or tuft of leaves. Generally, its **leaves** are a yellow-green and they are simply pinnate with two to four pairs of leaflets and a terminal leaflet. Occasionally there may be six pairs, or more rarely, up to eight pairs of leaflets. The leaflets all point forwards with the upper ones crowded and sometimes overlapping and the lower ones slightly spaced. The leaf blade is usually 2.5–2.6 cm long. Its **flowering stem** is 15–20 cm long, yellowish to greenish, the **flowers** are creamy, numerous and distributed along the stem in small clusters. Flowering usually occurs between December and February. *Aciphylla monroi* is confined to the South Island from Nelson and Marlborough to about the mountains of Mid Canterbury. It is fairly common on rock outcrops, in snow tussock–herbfield and in open, partly eroded and open snow tussock grassland.

It is probably more common on the drier eastern ranges. It ranges from 1100 to 1700 m. *Aciphylla similis* (lower photo) is a closely related species that is sometimes confused with *A. monroi*. It may be recognised by having **4–10 pairs of leaflets** that are wider-spreading than those of *A. monroi*. Sometimes its colour is not as yellowish as that species, being more greenish. Its **flowering stems** may be up to 30 or 40 cm long with the **flower heads** being clustered near its top or, more usually, distributed along the stem. It is confined to the South Island from just north of Lewis Pass to about the Two Thumb Range in South Canterbury but is rather erratic. It can be found in low-alpine regions on both sides of the main divide and generally favours the higher rainfall areas, ranging from 900 to 1100 m. It occurs in wet, mixed snow tussock–scrub and snow tussock–herbfield, and on rock outcrops.

APIACEAE

52

Aciphylla pinnatifida

This is a distinct and easily recognised species that normally grows in damp or wet areas. *Aciphylla pinnatifida* is also unusual in that it has a **rhizomatous growth habit**, something not shared by any other species of *Aciphylla*. The central rootstock puts forth underground rhizomes that form offset plants a short distance from the main rosette. Its **leaves** generally lie flat on the ground and are very distinctive; each is deeply divided to its yellow midrib. A pair of leaf-like **stipules** spread very widely from where the petiole joins the leaf blade, so that the leaf appears to be trifoliolate. These stipules are so leaf-like

that they are virtually indistinguishable from the actual leaf blade. The leaves also have prominent yellow midribs. When in **flower**, the difference between male and female plants is quite marked. The flowering stems of the female plant are 15–20 cm long and each cluster of flowers is more or less enclosed by a prominent orange, sheath-like bract about 3 cm by 2 cm, each bract sheath being terminated by a leaf-like spinous tip. *Aciphylla pinnatifida* usually flowers between December and January. It occurs in the higher mountain areas of western Otago, northern Southland and Fiordland, and ranges from 1100 to 1700 m. Its usual habitat is in wet places such as alongside streams, in wet depressions and on snow-banks.

APIACEAE

Kopoti is a herb that is quite variable and includes some rather large varieties that inhabit the low country as well as alpine forms that can be bewildering in their diversity. Some varieties are named but there are many alpine forms that have not yet been recognised. It is a **rosette herb** arising from a stout and long **taproot**, and usually its leaves are rather flattened on the ground. The rosettes of some alpine varieties are quite dwarfed and may be no more than 3 cm in diameter while others may be about 10 cm in diameter. Its **leaves** vary from green to a bronzed green and they all have paired **leaflets** that are rounded or fan-shaped. Each leaflet has toothed margins and the teeth are tipped with quite long bristle points. The branched flower stems of the kopoti may be up to 10 cm long (or on some of the larger varieties up to 50 cm). On male plants the numerous small, white **flowers** are produced in umbels or heads, the outermost ones being 3–4 cm across. On female plants the umbels are usually smaller and shorter with fewer flowers. The flowers are aromatic. Flowering usually occurs some-time between October and December. Kopoti occurs in the North and South islands from the Waikato and East Cape southwards, ranging from 450 to 1650 m. It inhabits grasslands and herbfields and also occurs on rock faces and prefers rather open habitats. The plant is very palatable to browsing animals and in some areas may be more common on inaccessible sites than in more open areas such as grasslands.

Anisotome haastii is a distinct and easily recognised species that occurs in the higher rainfall regions of the Southern Alps. It is a large, **tufted herb** that will form a clump up to 30 cm across. It has deep green, fern-like **leaves** that are two to three pinnate and much divided. The leaf blade is 10–25 cm long by 5–12 cm broad with narrow segments that end in fine, hair-like tips. Its **petioles** are 8–12 cm long and light green to purplish. The **flowering stems** may be up to 60 cm tall and have rather large umbels of numerous small, white flowers. There is not a great deal of difference between male and female plants. Generally, it can be seen in flower between the

months of October and February. This species is found in subalpine to low-alpine areas in the high-rainfall regions of the South Island, particularly about and west of the main divide. It is usually rare or absent from the drier mountain areas such as Marlborough and Central Otago. It ranges from 600 to 1520 m. *Anisotome haastii* frequently grows in herbfields, snow tussock–herbfields, areas of subalpine scrub, fellfields, rocky places, shingly ground and rock bluffs. It used to be heavily browsed by mammals such as sheep, deer and hares, but where they have been controlled or reduced it has once again been able to increase and become more common.

APIACEAE

Anisotome imbricata is one of the smaller species of *Anisotome* and also one of the more unusual ones. It grows on the high, rolling tops of the Otago mountains and is an excellent example of the way in which a plant responds to the challenges of its habitat. It forms small, **tight cushions** or hummocks, up to 20 cm high and 40 cm or more in diameter. From the outside the cushions have a silvery appearance with an underlying hint of brown that is caused by silvery and light brown hairs on the ends of the leaflet segments. Its **above-ground stems** become quite woody with age and they terminate

Anisotome imbricata var. *imbricata*

with small rosettes of leaves. The **blade of each leaf** is about 1 cm long with four to eight pairs of very close-set, overlapping pinnae. The pinnae have three to six teeth and have hairs at the tips of each segment. The **flowers** are on very short stems, more or less hidden among the leaves. There are two varieties of this species. Variety *imbricata* grows mainly in the cushion plant association of the tundra-like vegetation of the Central Otago mountains. It occurs in fellfields and on exposed ridges close to the upper limit of the snow tussock–herbfield, and it ranges from 1200 to 2000 m (see lighter shade on map). Variety *prostrata* usually has loose and more numerous creeping stems that give rise to open, green mats.

Anisotome imbricata var. *prostrata*

Its leaf hairs are brownish. This variety occurs from Marlborough and Nelson to Otago. In Central Otago it inhabits alpine bogs and wet depressions in snow-banks. It ranges from 1200 to 1800 m.

Anisotome pilifera

Anisotome pilifera is a stout and very handsome species. It has a strong, deeply descending **taproot** and is usually about 40 cm tall when not in flower. Its **petioles** are up to 10 cm long and are usually coloured purplish. It has bold and handsome **leathery leaves** but their colour varies according to the part of the country in which they occur. Those of the Nelson region are a bright, deep green with shiny upper surfaces, while the leaves of those from further south are a most

attractive glaucous colour. The **leaf segments** are generally rather broad, two- to three-lobed or coarsely toothed around their margins. Its **flower stems** are up to 60 cm tall with numerous small, white

flowers. Flowering generally takes place between November and March. This species is widespread in low- to high-alpine regions of the South Island from north-west Nelson and Mt Stokes in the Marlborough Sounds southwards, but in Otago it is absent from all but the most westerly ranges. It ranges from 1060 to 2130 m. *Anisotome pilifera* usually occurs on rock faces and bluffs where it grows in crevices and on ledges. It may also occur in more accessible sites in fellfields and herbfields but, in many areas, it has been much reduced by the browsing of introduced mammals.

APIACEAE

This is a summer-green herb that is another one of the remarkable plants which only inhabit the shingle screes along the eastern side of the Southern Alps. As with most other inhabitants of these inhospitable areas, it has thick and fleshy leaves that are a glaucous-grey colour that very well matches the colour of the scree. It is barely more than 10 cm high while the largest plants may be up to about 15 cm across. *Lignocarpa carnosula* arises from a rather long and twisting **rootstock** buried deeply in the shingle scree. From the top of the rootstock one or two shoots arise and make their way to the surface of the scree where they then develop as leaves. The **leaves** are deeply and finely divided into narrow, fleshy and rounded segments that are so closely placed that they greatly protect the plant from the desiccating effects of wind and insolation. Usually, the surface layer of the

scree does not move very quickly (unless disturbed by the passage of some large animal) so that the foliage is well able to cope with the gradual movement of the stones. The minute **flowers** are produced as small clusters within the shelter of the leaves. Their petals are whitish, creamy, or sometimes even pink has been recorded. It has been observed that after the seeds ripen the leaf breaks its connection with the rootstock and, as it dries, behaves as a kind of tumbleweed, being bowled across the scree to disperse its seeds as it goes. *Lignocarpa carnosula* occurs in low- to high-alpine areas from Nelson and Marlborough southwards to about the Rangitata Valley in South Canterbury, ranging from 900 to 2100 m. It is confined to the greywacke screes that are such a feature of the drier mountains along the eastern side of the Southern Alps.

Snowberry *Gaultheria depressa*

This species of snowberry comprises two varieties: variety *depressa* is the typical form and is distinguished by its leaves being almost rounded with prominent bristles around their more or less scalloped margins. Variety *novae-zelandiae* has narrower, more pointed leaves that are more obviously toothed and without persistent bristles. The snowberry is quite a small, **low-growing shrub** that is often only about 20 cm across and seldom more than 5 cm high. It has trailing or prostrate branches that form matted patches and root into the ground. The small **leaves** are about 5–10 mm long by 4–8 mm wide, and are rounded with small teeth around their margins. Their upper surfaces are deep green and they are paler beneath. Both varieties have small, bell-like, white **flowers** that are tucked into the leaf axils. The flowers are followed by relatively large, fleshy **fruits** up to about 1.5 cm in diameter. The fruits are formed by the calyx becoming swollen, large and succulent. The fruits vary in colour from white to pink or red and may be seen on the plants from January to April. In spite of glowing reports by some

authors extolling the virtues of their flavour they are actually rather tasteless. Early settlers in some southern districts used to make snowberry pies. In Otago *Gaultheria depressa* used to be known as nardoo berry and the southern Maori name for this species is tapuku. *Gaultheria depressa* var. *depressa* occurs in montane to low-alpine areas from the southern Tararua

Range in the North Island, and along the main divide and around the Dunedin district in the South Island, ranging from 500 to 1500 m (see lighter shading on map). It grows in subalpine forest and scrub, open areas in snow tussock grassland and herbfield, and rock outcrops. Variety *novae-zelandiae* is found in similar sites but is widespread in mountain regions from the Volcanic Plateau of the North Island and southwards through the South Island. It ranges from lowland to 1500 m. Along the eastern side of the South Island white-fruited forms appear to be most common.

Acrothamnus colensoi, also known as *Cyathodes colensoi*, is a small **shrub** growing to about 40 cm tall by 60 cm or so across but, occasionally, it may be much wider spreading. Older and larger plants form broad, almost flattened patches. It usually has rather dense growth and the bark of its **branches** is grey. Its small **leaves** have a distinctive shape, being oblong and suddenly narrowed at the tip to a minute point. Their upper surfaces are dark green or greyish while beneath they are pale greyish and are marked by about five distinct parallel veins. The hairy **margins** are rolled slightly downwards and inwards. The small, white **flowers** are about 4–5 mm long, tubular and are produced in clusters of two to five at the tips of the branchlets. The tube is hairy in its upper part. The tip of the tube has five lobes that are densely hairy on their upper surfaces and the flowers are distinctly honey-scented. Often the flowers function only as male or female, even though they all have male and female parts. Only those that are functionally female bear fruit. Flowering usually occurs between December and January. The **fruits** are 4–5 mm in diameter and on different plants may range from red to deep crimson to pink to white. The various colours may often be seen growing in proximity to each other. Fruits are present on the plants between January and April. *Acrothamnus colensoi* is found in montane to low-alpine regions in both the North and South islands from the Volcanic Plateau, the Kaimanawa Mountains and the Ruahine Range southwards. In the South Island it is mainly on the drier mountains of eastern Nelson and Marlborough southwards to Canterbury and then it may be local in Otago and rather rare in northern Southland. It ranges from 600 to 1600 m. It usually grows on well-drained sites in mixed tussock–scrub, snow tussock grassland, rock outcrops and fellfield.

Patotara *Leucopogon fraseri*

Patotara is a small, **low-growing shrub**, with leafy branchlets, that may form quite extensive patches. It extends its territory by **stems** that creep along, or just below, the surface of the ground. Its wiry **aerial stems** are erect, 5–15 cm tall, and clothed with foliage for most of their length. It is easily recognised by its **leaves** that are hard-textured and have pungently pointed tips. They are 4–9 mm long by 1–3 mm wide. Their upper surfaces are deep green, greyish or yellowish to bronzy green. The white **flowers** are produced along the upper part of the stem from the leaf axils. They

are strongly honey-scented, 1–1.3 cm long and have five distinctly bearded lobes at the mouth of the tube. Flowering usually occurs between September and January. The **fruit** is an orange edible berry, 8–9 mm long. It is sweetish but may have a somewhat resinous taste. The style of the flower is persistent and remains attached to the end of the fruit like a whisker, which is another distinctive feature. Fruit is usually present between November and March. Because the patotara is not palatable to browsing stock it is one native species that has benefited from European settlement. It is found in the North, South and Stewart islands in coastal to low-alpine regions throughout. It is usually common in the low tussock grasslands of the drier regions east of the main divide and fellfields, but may also be present in higher rainfall areas if very well-drained and open such as moraine gravels and stabilised riverbeds. It ranges from coastal areas to 1600 m.

This is a dwarf, much-branched shrub that forms **dense patches or low hummocks** up to 4 cm thick and 50 cm or more across. Because of its habit of producing both flowers and fruits at the same time, it is generally easily recognised. Its small **leaves** are 3–5 mm long by 1–2 mm wide and they are a distinctive deep green or bluish green. The small, whitish **flowers** are borne singly near the tips of the branchlets and they are rather similar to those of *Acrothamnus colensoi* (see p. 60). They are also honey-scented. Its berry-like **fruits** are 5–6 mm in diameter and are bright red but they are hollow, not fleshy, and contain five or more small nutlets. Its hollow fruits help to distinguish this species from other species of a similar appearance. The fruits take up to two seasons to ripen; by the first autumn they are small and green, and then by the following summer grow to maturity. It is quite usual for plants of *Pentachondra pumila* to have new season's flowers as well as displaying ripe fruits from the previous season all at the same time. It is distributed widely in the North, South and Stewart islands, from Mt Moehau on the Coromandel Peninsula southwards, and it is found in subalpine to low-alpine areas. It is rare north of East Cape but is very common throughout the South and Stewart islands. It occurs in a variety of habitats including cushion bogs, open snow tussock grasslands, herbfields and herb moor. It also grows in exposed or rocky sites or in poorly drained peaty areas. It ranges from 600 to 1500 m.

This is a prostrate and much-branched, rigid **shrub** usually no more than 50 cm tall and spreading perhaps to about 50 or 60 cm. It is easily recognised by its small and stiff, grass-like leaves, usually of a reddish or brownish colour, and its stiff and rigid branches. The **bark** of the older branches is grey and somewhat rough. Its **branches** and its branchlets are usually short. The quite short **leaves** are generally crowded towards the tips of the branchlets and more or less erect to spreading. They are 7–8 mm long and from their broad sheathing bases they then suddenly contract to a very narrow blade with a blunt tip. The colour of the leaves depends upon locality and

they can vary from green or yellowish green to a brown or bronze or reddish brown. Usually, plants growing in the more exposed situations have reddish brown or brownish leaves, particularly as conditions become cooler. The white to creamy-white **flowers** are terminal, solitary at the tips of the branchlets, and they are sweetly scented. The flowers are produced during summer. *Dracophyllum pronum* occurs throughout the higher mountain regions of much of the South Island in low-alpine to high-alpine regions. It usually grows in snow tussock grassland and herbfield, especially in depleted areas where the cover is thin and weak. It also occurs on exposed ridges, in rocky places and in fellfield. It ranges from 800 to 1800 m.

MYRSINACEAE

Myrsine nummularia is usually a rather easily recognised little **shrub** of **prostrate or creeping** habit and with small, almost rounded leaves that are alternate along its branchlets. It can form patches up to 50 cm or more across. In more open situations it will form quite dense masses, but when creeping through other vegetation its growth may appear to be rather sparse. Its thin **stems** are pale brown to reddish brown. The rather thick **leaves** are about 4–10 mm in diameter and their upper surfaces are light green to bronze, usually with a dark patch near the base of the leaf. The pale undersurface of the leaf is distinctly dotted with small oil glands. The **flowers**

are either solitary or borne in few-flowered clusters; they are quite small and the male and female flowers are produced on separate plants. Flowering usually occurs between October and February. Its globose **fruits** are 5–6.5 mm in diameter; a lovely violet colour, but they are often hidden among the leaves. They take 12 months to ripen so that when the plant flowers the fruits displayed at the same time are the result of the previous season's flowering. Fruits may be seen on the plants between December and May. *Myrsine nummularia* is found in the North, South and Stewart islands. It is fairly widespread in montane to low-alpine regions from East Cape southwards. It does not occur on Mt Taranaki and is often local elsewhere. This species usually grows in sheltered places in tussock grassland, tussock–low scrub, snow tussock–herbfield, open alpine shrubland and around rocky outcrops. It generally ranges from 600 to 1520 m, but on Stewart Island it descends almost to sea level.

Coprosma 'brunnea'

This is a low-growing or sprawling **shrub** some 30–40 cm high that may be up to 50–60 cm across, but sometimes it may be larger. Its slender, brown or reddish brown **branches and branchlets** are flexible, zigzagging and interlacing so that it forms a rather tangled mass. It has very small **leaves** that are either in pairs or small clusters. The leaves are 7–11 mm long by 1–1.5 mm wide and they are usually a dark, brownish green to dark green. Its **flowers** are quite inconspicuous, although the flowers of **male plants** have prominent anthers and may be more noticeable. **Female plants** have quite conspicuous **berries**, usually of an attractive sky-blue, and often flecked with darker blue, although the colour may vary. The fruits usually ripen during February. This species is now considered to be an alpine variant of the shore coprosma (*C. acerosa*), although some people still prefer to classify it as a separate species. It occurs throughout the North, South and Stewart islands, from Mt Taranaki and the Volcanic Plateau southwards. It ascends to 1500m, ranging from lowland to low-alpine regions and grows on stony river and streambeds and grassy river terraces. It may also be found in subalpine scrub and sometimes it may occur in open, rocky areas in tussock–herbfield.

RUBIACEAE

This is a small species of *Coprosma* that creeps along the ground to form **small patches**, though its growth is often intermingled with other vegetation. When not entangled with other vegetation it may form patches up to 60 cm or more across. Its **growth** is quite prostrate, seldom arising more than a centimetre or two in height. The **leaves** are often clustered on short branchlets; the leaf blade is 4–10 mm long by 3–4 mm wide. Leaves are elliptic, thick and almost fleshy and their upper surface is bright green and shiny. While its **flowers** are inconspicuous, the pale anthers of the male flowers and the long, whitish, branched styles of the female flowers normally stand out from among the surrounding matted vegetation in which it grows. For such a small plant its orange **fruits** are relatively large, being about 6–10 mm in diameter. It flowers during summer and the fruits ripen about March. *Coprosma perpusilla* occurs in subalpine to high-alpine regions of the North, South and Stewart islands from near East Cape southwards, ranging from 800 to 1900 m. It is often quite widespread in a variety of habitats in tussock–herbfield, fellfields, cushion bogs and on snow-banks, usually where the soil is permanently moist. It also occurs on the Auckland, Campbell, Antipodes and Macquarie islands. It was formerly known as *Coprosma pumila* but that species is now considered to be confined to Australia.

Everlasting daisy *Anaphalioides bellidioides*

The everlasting daisy is an easily recognised species that usually forms trailing or creeping **patches**. It has almost woody **stems** that root into the ground and are quite thin. They usually become upright towards their tips. Its small **leaves** may be quite variable, their upper surfaces varying from green to quite silvery with the undersurfaces being whitish. They are closely placed and spreading, 5–8 mm long by 3–5 mm wide, and usually broadly oval-shaped. Its **flowers** are produced singly on short stalks, up to 10 cm long, that arise from the tips of the branchlets. The flower heads, including the white,

papery petal-like bracts (modified leaves) that surround them, are 1.5–3 cm in diameter. It usually flowers between October and February, although the old, partially dried flower heads often give the impression that it is still in flower beyond its normal season. Everlasting daisy is a very common plant in the alpine vegetation and may be found in quite a variety of situations. It occurs in the North, South and Stewart islands as well as also extending to the Chatham, Antipodes, Auckland and Campbell islands. On mainland New Zealand it is found from East Cape and Mt Taranaki southwards, growing in lowland to low-alpine areas, from sea level to 1600 m. It occurs in scrub, tussock grasslands, herbfields, riverbeds, stony places, road banks and rocky outcrops.

Brachyglottis bellidioides

This alpine daisy belongs to quite a variable species that can exist in a bewildering number of forms. Not infrequently, its foliage is tucked deeply into the surrounding vegetation so that it can be quite inconspicuous, except perhaps for its yellow flower heads, peeping out from the vegetation, that are often the only indication of its presence. It is a small **rosette herb** that arises from a stout, often unbranched rootstock. The **leaves** are usually closely flattened to the ground or pressed tightly against the vegetation in which it grows. They are on short, hairy stalks about 1–2.5 cm long. The **leaf blades** are 1–5 cm long, oblong to more or less rounded with a blunt or slightly pointed

tip and rounded at the base or narrowing to the leaf stalk. The upper surface of the blade is medium to deep green with numerous to few short hairs while the undersurface is paler and has few or no hairs. This latter character enables it to be distinguished from the similar *Brachyglottis lagopus* (described on p. 72). The usually unbranched **flower stems** are up to 30 cm tall and are covered with cottony hairs. They carry just one flower head, 2–3 cm in diameter. The **flower head** has bright yellow ray florets (petals) and the central disc florets are orange. Flowering is usually between October and March. *Brachyglottis bellidioides* is fairly widespread throughout mountain regions of the South and Stewart islands, ranging from 300 to 1800 m. It occurs in montane to high-alpine areas, being found in tussock grasslands, snow tussock–herbfield, open shrublands, herbfields and on rocky bluffs. In the drier grassland areas it more commonly grows in moist and sheltered situations.

ASTERACEAE

This is a complex species that is found in alpine shrublands in both the North and South islands. It is a **shrub** around 1–3 m tall and has stout, spreading **branches**. Its very thick and **leathery leaves** are 5–9 cm long by 3–5 cm wide, their upper surfaces are dark green and shiny and the undersurfaces are usually whitish or a dirty buff colour. The **flowers** are borne in terminal panicles that are up to about 15 cm long. The yellowish **flower heads** are more or less bell-shaped,

about 8 mm in diameter, and they have only disc florets but no ray florets (petals). Flowering usually occurs between December and February. The North Island form of this shrub is generally known

by the name of *Brachyglottis elaeagnifolia* and it is found in alpine forest and scrub from the mountains of East Cape southwards. In the South Island it is sometimes referred to as *Brachyglottis bennettii* and is found in lowland to higher montane forest and scrub, in the higher rainfall regions, from north-western Nelson southwards. It ranges from lowland areas to 1400 m. Opinion is divided as to whether *B. bennettii* should be known by that name or just regarded as a form of the North Island *B. elaeagnifolia*. *B. bennettii* does have thinner and less leathery leaves than its North Island relation.

ASTERACEAE

One of the native shrub daisies, this is a small **bush** that grows from about 30 cm to 1 m tall. It exists in two varieties: **variety** *bidwillii*, which is the typical form (confined to the North Island) and **variety** *viridis*, which occurs only in the mountains of the South Island. *Brachyglottis bidwillii* is recognised by its compact habit of growth and its small, thick and leathery **leaves** that are wider towards their blunt tips. Their upper surfaces are a shiny, dark green and the whitish or buff undersurfaces have a felted appearance. Its **flower heads** are a creamy colour and they have disc florets only, lacking the petal-like ray florets that one might expect of a daisy. They are 7–15 mm in diameter and are produced in tight clusters on branched flowering stems. Flowering usually takes place between January and March, according to altitude. *Brachyglottis bidwillii* var. *viridis* is distinguished by being a taller plant, having more slender branchlets than *Brachyglottis bidwillii* var. *bidwillii*, which has larger and slightly less leathery leaves. The typical form is fairly widespread in montane to subalpine areas of the North Island from the mountains of East Cape to Lake Taupo and southwards to Cook Strait. On the central volcanoes it extends into the fellfield. *Brachyglottis bidwillii* var. *viridis* is distributed from the mountains of Nelson and Marlborough southwards to about the Rakaia Valley in Mid Canterbury, ranging from 800 to 1700 m. It occurs in low subalpine scrub, snow tussock–herbfield and on rocky outcrops. When in scrub it will attain its maximum size but at higher altitudes and in more exposed situations it is usually much more compact. Plants growing on rocky outcrops are frequently dwarfed to no more than 20 or 30 cm high.

Brachyglottis bidwillii var. *viridis*

70

In general appearance this species is similar to larger forms of *Brachyglottis bellidioides* (see p. 68) except that it does not have its leaves flattened to the ground and they tend to be more erect. It is also distinguished and easily recognised because of the whitish colour of its leaves. It is a **tufted herb** that has a branched rootstock and it usually forms small clumps up to about 15 or 20 cm across. Its **leaves** form rosettes, at the ends of the branches, and they have petioles 3–15 cm long. The leaf blades are oblong to ovate, usually rather leathery, and their upper surfaces are densely clad with a white tomentum, but with age the tomentum gradually wears off. The margins of the leaves are more or less smooth or they may have obscurely rounded teeth, but they are more obvious than those on *Brachyglottis bellidioides*. The tip of the leaf is blunt while the base is rounded to somewhat heart-shaped. The **flowering stem** is up to 35 cm tall, rather slender and branched. Its **flower heads** are 2–4.5 cm

in diameter and the ray florets are yellow. Flowering usually occurs between December and February. *Brachyglottis haastii* is found in the South Island where it occurs in montane to low-alpine regions, usually in the drier mountain and hilly areas from south-eastern Nelson to western Marlborough and then southwards to the Otago lakes district. The species may be rather uncommon except in the Mt Cook district. It ranges from 300 to 1500 m, and grows in fairly dry grassland, open shrubby areas or, more commonly, on rocky outcrops. It often grows on limestone but is by no means confined to such conditions.

Brachyglottis lagopus

As with *Brachyglottis bellidioides*, this species is quite variable and exists in a number of different forms. Some lowland forms are large, tufted herbs with leaves that are up to 20 cm or more long, while alpine forms may have only single rosettes with leaves no more than 4 cm long. Proportionate to their size, they all have thick **rootstocks**, the upper parts of which are covered with woolly brown hairs. In general appearance, alpine forms of this species are not too dissimilar to *Brachyglottis*

bellidioides, which differs by having no hairs or just a few sparse hairs on the undersurfaces of its leaves. The **leaves** of *Brachyglottis lagopus* are spreading and mostly lie flat on the ground, their **petioles** are 1–3 cm long and the leathery leaf blade is somewhat oblong to ovate or more or less rounded, 2.5–5 cm long by 2–4 cm wide. Their upper

surfaces are deep green with a rather crinkled texture along the veins; the undersurfaces are densely covered by a soft, white or buff, woolly **tomentum**. Its **flower stems** are up to 35 cm tall (those of alpine forms are usually much shorter) and may be unbranched, with just a single flower, or branched with several flowers. The bright yellow **flower heads** are 2–3 cm in diameter, and flowering is usually between November and January. *Brachyglottis lagopus* is fairly widespread throughout the North and South islands from the Taupo region and the Ruahine Range southwards to about South Canterbury. It is found in lowland (but only in some localities such as coastal Marlborough and Banks Peninsula) to high-alpine regions, ranging from near sea level to 1500 m. It can be rather common in open shrubland, tussock grasslands and snow tussock–herbfield, particularly on rocky sites.

Celmisia armstrongii

Celmisia armstrongii is a very handsome species of mountain daisy and is one that is most easily recognised. It is one of the larger **rosette-forming** species and frequently forms just the one rosette but older plants may branch and form clumps of several rosettes. It is easily distinguished by its narrow and rigid sword-like **leaves** that radiate stiffly from around the rosette. They are 20–35 cm long and no more than 1–2 cm wide, and the upper surfaces of the leaves are distinctively coloured a bronzy green with a yellow band along either side of the midrib. The undersurfaces of the leaves are white or buff coloured. Its rather stout **flower stems** are up to 25 cm long and, while the flower heads are not overly large, they are 4–5 cm in diameter. It usually flowers between December and January. *Celmisia armstrongii* is found in subalpine to low-alpine regions of the South Island, being found in the higher rainfall regions about and west of the main divide from Nelson to the Humboldt Mountains in western Otago, ranging from 800 to 1500 m. It can be quite prominent in snow tussock–herbfield.

ASTERACEAE

As well as being quite a common species, *Celmisia discolor* is also rather variable. It is closely related to *Celmisia incana*, and some forms are not always easily distinguished. Generally, *C. discolor* is distinguished from *C. incana* by its usually narrower leaves, and the hairy covering on their upper surfaces being quite thin so that the leaves have a greenish appearance rather than being pale grey or whitish as in *C. incana*. It is a **prostrate or sprawling** species that will sometimes form rather open patches up to a metre across. Generally, the **old stems** are covered with the remains of its old leaves. The **leaves** are

crowded towards the tips of the branchlets and are 2–4 cm long by 8–12 mm wide. They are more or less ovate and their upper surfaces are clad with a dense covering of soft, flattened hairs that gives them a greenish grey or slightly silvery appearance. Their undersurfaces are densely white. Its **flower stems** are 10–15 cm long and the white **flower heads** are 2–3 cm in diameter. It usually flowers between December and January. *Celmisia discolor* occurs in subalpine to low-alpine areas of the mountains of the South Island, ranging from 1000 to 1700 m. It is not common on the drier eastern ranges but otherwise occurs throughout. South of Westland it becomes rare and local.

This is a much-branched trailing or sprawling **sub-shrub** that forms rather dense to loose patches that, in some cases, may be up to a metre or more across. Its woody **stems** creep over the ground, rooting in as they grow, and their growing tips turn more or less upwards. The **leaves** are the most characteristic feature of this species. They are densely clad with white hairs; their undersurfaces are similar. Thus the clumps or patches of this plant are quite obvious and usually stand out for quite some distance. Its leaves are 2.5–4 cm long by 12–15 mm wide, more or less ovate to oblong and widest towards their tips, which are blunt or slightly pointed. The **flower stem** is rather slender, about 10–12 cm long and usually green or purplish. Its **flower heads** are 2.5–4 cm in diameter; they have white ray florets and yellow centres. Flowering usually occurs between late October and December. The similar *Celmisia discolor* (see previous entry) is a variable species that occurs in the wetter mountain regions, and some forms of it could be confused with *C. incana*. The leaves of *C. discolor* usually have green or greenish upper surfaces. *Celmisia incana* occurs in the North and South islands from Mt Moehau, near the tip of the Coromandel Peninsula, and Mt Hikurangi, near East Cape, southwards through the central Volcanic Plateau to the drier eastern ranges of Nelson, Marlborough and North Canterbury. *Celmisia incana* usually occurs in open grasslands, herbfields, rocky places and open tussock–scrublands, ranging from 900 to 2000 m. A very similar species is *Celmisia allanii*, which occurs in the mountains of Nelson and north-western Canterbury.

Mountain daisy *Celmisia semicordata*

Celmisia semicordata is an outstanding species that cannot fail to be noticed. It forms large, tufted **rosettes** of silvery, sword-like leaves that stand out from the surrounding alpine vegetation and it is generally regarded as the most magnificent of the mountain daisies. It usually forms a single rosette, but older plants normally have several **rosettes** so that they form quite large clumps. Its stiff and leathery **leaves** are 30–60 cm long by 4–10 cm wide and their upper surfaces are clad with a dense covering of soft, flattened hairs, giving it a silver or silvery-green colour. This covering protects the leaf against the damaging effects of the elements. The undersurfaces of the leaves are white. Its **flowering stems** are stout and up to 50 cm long. The **flower heads** are 4–10 cm in diameter; the largest is almost the size of a small saucer. Flowering occurs during December and January, the exact period depending on altitude. This mountain daisy is confined to the South Island, occurring in the higher rainfall regions from Nelson and northern Westland to South Canterbury and then to northern Fiordland, ranging from 600 to 1400 m. It becomes more common further south, as well as near to and west of the main divide. As well as the typical form two subspecies occur in the far south: subspecies *stricta* has much narrower leaves and occurs on the mountains of northern Southland (see lighter shading on map); and subspecies *aurigens* has leaves of varying shades of bronze or gold. It is confined to the mountain ranges of eastern and south-Central Otago (see brown shading on map). Mountain daisy is also known as cotton plant and tikumu.

Celmisia sessiliflora

This is a very compact little species of *Celmisia* but is one that can be quite noticeable out in the field. Although it does not appear so, it is actually classed as a **subshrub**. *Celmisia sessiliflora* forms dense, silvery, **low hummocky cushions** that may be from 20 cm to a metre or more in diameter and up to 10 cm high. Its rigid **leaves** are clustered around the tips of its branchlets to form dense rosettes. They are silvery-white or greenish white and are 1–2.5 cm long by 1.5–2.5 mm wide. Its white **flower heads** are 1.5–3 cm in diameter and are sunk among the tips of the rosettes. The epithet *sessiliflora* means that the flower heads have no stalks but in fact they do have very short stalks that

elongate and become more obvious as the seeds ripen. Flowering usually occurs between December and January but may occasionally be later. *Celmisia sessiliflora* is found in subalpine to high-alpine regions throughout the South and Stewart islands, ranging from 700 to 1800 m. It may be seen in a variety of situations from low tussock grasslands to short snow grass, herbfields, fellfields and cushion bogs. This last habitat appears to be more common in southern regions. It is an easily recognised species, often forming extensive patches in open situations such as on dry ridges as well as in permanently damp places.

Common cotton plant *Celmisia spectabilis*

Due to the past actions of sheep grazing and the former frequent burning of grasslands this is now one of the commonest and most widespread species of the celmisias. Its unpalatability to mammalian browsers as well as its ability to withstand frequent grass fires ensured that it became a prominent member of mountain grasslands. It is a rather robust **rosette herb** that either grows as a single rosette or forms clumps of several rosettes. Older plants may be up to a metre in diameter. The lance-shaped **leaves** are 5–20 cm long by 1–2 cm wide and their upper surfaces are bright, shiny green while their undersurfaces have a dense but loose covering of whitish- or buff-coloured woolly hairs. The margins of the leaves are usually slightly rolled downwards. These characters make it easy to identify this species. The leaves of older plants slowly decay to form a somewhat peaty mass that usually remains damp and they, along with the rather long leaf stalks that closely overlap to form a false stem, help to protect the yet-to-emerge young leaves from the effects of fire. The stout **flower stems** of the common cotton plant are 20–25 cm long and each stem carries a single **flower head**, 3–5 cm in diameter. Flowering usually occurs during early to mid summer. Common cotton plant occurs in montane to low-alpine regions throughout the North and South islands, ranging from 300 to 1700 m. In the North Island it is distributed from Mt Hikurangi and the central volcanoes southwards. In the South Island it is found mostly in the drier tussock grasslands east of the main divide to about as far south as the Rakaia and Mathias rivers. It usually grows in red tussock grasslands and fellfields in the central North Island and snow tussock–herbfield in the higher rainfall areas of the Tararua Range but is absent from the eastern Wairarapa area. In the South Island it often occurs in the higher rainfall regions of north-west Nelson to Arthur's Pass. In the Nelson region plants of this species are considerably smaller than those from other regions while those from South Canterbury are large and much more robust. It is also known as puharetaiko.

Celmisia traversii

This is a magnificent species that is easily recognised, even when not in flower. It is usually a **stout herb**, 35–50 cm tall, that grows as a single rosette or as a clump of multiple **rosettes**. Its **leaves** are 15–30 cm long by 4–6 cm wide and are distinguished by their rather dull or somewhat shiny, deep green upper surfaces and particularly by their undersurfaces being densely clad with a soft covering of pale to rich rusty-brown, velvety hairs that also prominently fringe their margins. The **leaf stalks** are a dark purple colour which also extends for a short distance onto the midrib. The base of the leaf blade abruptly narrows to the leaf stalk. The **flower stem** is 20–50 cm tall and like the undersurfaces of the leaves is densely clad with soft, rusty-brown hairs. Its **flower heads** are 4–6 cm

in diameter and are surrounded by narrow bracts clad with soft, rusty-brown hairs. Flowering usually occurs between December and January. *Celmisia traversii* is found in the higher rainfall areas of the low-alpine regions of the South Island, ranging from 900 to 1600 m. It extends from north-western Nelson to western Marlborough southwards to north-western Canterbury; there is then a considerable gap of about 500 km before it reappears in southern Fiordland and western Southland. It usually occurs in permanently moist snow tussock grassland to snow tussock–herbfield.

ASTERACEAE

Craspedia is a genus that needs a great deal of study in order to resolve just how many species it actually contains. One species that is quite likely to be reasonably noticeable is the woolly head. It is a summer-green herb growing 6–10 cm tall, and it forms **rosettes** of rather soft **leaves** that are greyish-white because they are covered with hairs tightly pressed against their surfaces. A somewhat similar species (*Craspedia incana*) has snowy-white leaves, and may be found on some shingle screes; its leaves have a much woollier appearance. The leaves of the woolly head are 5–10 cm long by 1.5–2.5 cm wide and they are widest above their middles. They narrow gradually to

rather broad, flat petioles. With age the hairs may gradually wear off the leaves so that they are no longer greyish-white. Its **flower stems** are 6–30 cm tall and are topped with a single yellow **flower head** about 2 cm in diameter. In shape they are hemispherical or slightly top-shaped. Flowering is usually about November to December or January. The woolly head is confined to the South Island and is found in montane to high-alpine regions, from 500 to 1900 m, where it occurs on the drier mountains east of the main divide. It usually grows in tussock grasslands, herbfields, fellfields and in cushion communities.

Dolichoglottis lyallii

This is a beautiful summer-green herb that is quite distinct when seen among the usual run of mountain vegetation. It has a stout, more or less buried **rootstock** that is clad with the remains of the old leaf bases. From the top of the rootstock sprouts a mass of bright green, narrow, rather fleshy, grass-like **leaves** up to 30 cm long by up to 1 cm wide. In the south, the leaves of some forms are often rather purplish. The **flower stems** may be up to 50 cm long, are often inclined outwards rather than standing erect, and usually have a broad corymb with numerous flower heads. The **flower heads** are 4–5 cm in diameter and the plant is most attractive when seen flowering alongside a small mountain stream. It usually flowers during December and February. *Dolichoglottis lyallii* occurs in montane to high-alpine regions of the South and Stewart islands, ranging from 600 to 1800 m. On the drier eastern ranges it is usually local or absent, preferring higher rainfall regions. It mostly prefers permanently wet situations on shady slopes as well as cool, well-lit situations along open streambeds, alongside waterfalls and avalanche chutes, and it will also occur in shrublands. It not infrequently descends into open situations in forests. At higher altitudes it may be found in seepage areas, bogs, wet hollows, alongside streams and on shady rock outcrops.

Dolichoglottis scorzoneroides

While this species is somewhat similar to *Dolichoglottis lyallii* (see previous entry), it can be easily recognised by its broader and stiffer foliage and its larger, white flowers that are held on erect stems. It is a most beautiful species that graces many a mountain meadow. As with *D. lyallii*, it is summer-green and commences to die down with the onset of autumn. In suitable sites it is quite lush and will grow to 50 cm or more tall. Its stout **rootstock** is clad with the remains of the old leaf bases. The rather fleshy **leaves** are up to 20 cm long by 2 cm wide, their upper surfaces medium to deep green and usually clad with a few hairs. Its **flower stems** are stiffly erect and carry a

broad, terminal corymb. The individual **flower heads** are 4–6 cm in diameter and they have white ray florets. Flowering occurs between December and January. Where *D. scorzoneroides* and *D. lyallii* grow in close proximity to each other they will hybridise to produce some very attractive plants with flowers varying from white to cream, yellow and almost a salmon colour. *Dolichoglottis scorzoneroides* occurs in the South and Stewart islands in low-alpine to high-alpine areas, from 900 to 1700 m. It favours the wetter mountain regions, usually common about and west of the main divide. It normally occurs in herbfields, on rocky bluffs and in moist fellfields. In southern regions it also occurs on snow-banks.

Vegetable sheep *Haastia pulvinaris*

This is one of the most distinctive and remarkable plants to occur in the high mountains of the South Island. The name vegetable sheep was originally coined by early high-country musterers and was first recorded in 1899. The species is a **shrub** that forms quite large tight masses or cushions 30 to 60 cm high by as much as 2 m across. The **branches and branchlets** are very tightly packed so that only the growing tips are exposed; the rest is inside the cushion. The **leaves** closely overlap each other and, as the cushion increases in size, the leaves die but remain attached to the branches so that they only slowly decay, to form an almost peaty mass that always remains moist, even under the driest of conditions. The **outside leaves** are covered with woolly, buff-coloured hairs that give the plant its sheep-like appearance. From a distance they present very much the appearance of a flock of sheep browsing on the mountainside. Its yellow **flower heads** are about 1–1.5 cm in diameter and are sunk among the growing tips of the branchlets. Flowering usually occurs during December and January. Although vegetable sheep can be observed from some distance they occur quite high up so that to see them usually requires some climbing. But it can be very rewarding to study the plants close up. The species occurs in low- to high-alpine regions on the drier mountains from Marlborough to south-eastern Nelson, ranging from 1300 to 1900 m. It usually grows on shattered rock in fellfields or on stable scree slopes where there is little downward movement of the debris.

ASTERACEAE

This species is a relative of the vegetable sheep but is so completely different that people could be forgiven for believing that it is no relation whatsoever. It usually forms **loose patches** seldom more than 30 cm across with just a few widely spaced branches. Its **branches** are deeply buried in the loose debris and are no more than 3–5 cm tall, or less, at their erect tips. The spreading **leaves** are 3.5 cm long by 1.5 cm wide; they overlap and are covered with white, woolly hairs, and their tips are slightly pointed. The upper surface of the leaf is characterised by being furrowed by a series of parallel veins (visible through the hairs) that give them a very distinct appearance. The whitish **flower heads** are about 3 cm in diameter and are partially sunk among the leaves at the tip of the branch. Flowering occurs

between December and January. This is another scree plant that requires a bit of effort if it is to be viewed in the field. It occurs in low- to high-alpine regions of the South Island, from Nelson to Southland and Fiordland, ranging from 1300 to 2000 m, and may be quite widespread throughout. It normally grows in very stony fellfields, stable talus and stable screes.

Coral shrub *Helichrysum coralloides*

The coral shrub is well named, as it is very distinct and appears to resemble a large piece of coral. It is a stout and compact **shrub** growing about 20–60 cm tall and large plants may be up to almost 1 m across, although often smaller. It has rather **stout branches** 7.5–10 cm in diameter and they are often more or less prostrate but ascend at their tips. The **leaves** are very closely flattened to the stems and are very close set. Each leaf is 5–7 mm long by about 2.5 mm wide and they are positioned on the stem so that their **outer surfaces**, exposed to the weather, are bright green and shiny, and the inner surfaces, covered with **white woolly hairs**, face inwards towards the stem, thus being protected against excessive transpiration and the harsh conditions of its habitat. The branches and the spaces around the leaves are covered with dense, white woolly hairs so that it has the unusual appearance of a mosaic of deep, shiny green on a soft, cottony white background. Its solitary **flower heads** are whitish or yellowish, about 5–8 mm in diameter, and are produced from the tips of the branchlets. Flowering is normally between December and January. The coral shrub is confined to the South Island in montane to subalpine regions east of the main divide from the upper Awatere River in Marlborough to Mt Percival, near Hanmer Springs, in North Canterbury. It ranges from 900 to 1520 m. It usually grows on rocky places and in fellfields.

Helichrysum intermedium

In some respects this species resembles a much smaller version of the coral shrub (see previous entry). It is also a densely branched shrub and, to the uninitiated, can be mistaken for a species of 'whipcord' hebe, but nothing could be further from the truth. It forms a compact **shrub**, 15–40 cm high by 30 to perhaps 60 cm or more across. The **branchlets** are 2–4 mm in diameter and usually a deep, shiny green because of the scale-like leaves that enclose the branchlet. Each **leaf** is tightly flattened against the stem and sharply outlined by the white, woolly hairs that are also present on the inner surfaces of the leaves. Its solitary **flower heads** are creamy to almost white, 3–4 mm in diameter, and produced from the tips of the branchlets. Flowering usually occurs between December and February.

Helichrysum intermedium is common throughout subalpine to high-alpine regions of the South Island, ranging from 800 to 1600 m. It is mainly confined to rock outcrops and rocky places where it grows in crevices and similar situations. Plants on more shady rock faces often form more or less trailing masses up to a metre long. *Helichrysum intermedium* can be rather variable and one variety, *H. intermedium* var. *tumidum*, occurs at Cape Saunders on the Otago Peninsula. A related and similar species, *H. parvifolium*, occurs in Nelson, Marlborough and North Canterbury. It has finer branchlets and grows as a more erect shrub with bright citron-coloured flower heads.

Leptinella atrata

Leptinella atrata is another of those unique plants that has its habitat on the great shingle screes of the South Island's eastern mountains. It is one of about 21 different species of plants that inhabit the harsh and extraordinary environment experienced on those screes. As with most of the other scree plants, the colour of its above-ground parts blends very well with its surroundings and, particularly when it is not in flower, it is not always immediately noticeable. It is a fleshy herb with **creeping rhizomes** that are quite deeply buried in the scree. The rhizomes put forth **branching stems**, each of which terminates in a tuft of leaves just above the scree surface. Its greyish or grey-green **leaves** are 1.5–3 cm long, finely divided, with a fern-like appearance, and their margins are usually tinted reddish. The erect **flower stems** are 3–6 cm long, each bearing a single, dome-like flower head. Its **flower heads** are quite striking, being such a deep maroon colour that they appear to be black. They are 1–1.5 cm in diameter and when the outer florets open they display a circle of golden yellow anthers. Flowering usually occurs during December and January. *Leptinella atrata* is found on the drier mountains along the eastern side of the South Island from Marlborough to north Otago, ranging from 1000 to 2000 m. It is fairly widespread and while it usually favours the more stable scree slopes it can be relatively common on those that are more mobile. A less common variety (variety *luteola*) occurs in a few locations on the mountains of Marlborough and North Canterbury. It has greenish foliage and pale yellow flower heads. There is also a distinct species (*Leptinella dendyi*) that is found on mountains from Marlborough to Mt Hutt, in Mid Canterbury. It is a slightly larger plant with soft, yellow flower heads that are up to 2 cm in diameter.

Leptinella pyrethrifolia

This is a rather variable species with a creeping habit of growth that may be quite wide-spreading, up to a metre across, or it may form a smaller and sparser kind of plant. Its branched **rootstock** roots into the ground as it grows. It is purplish when young, becoming green with age, and the tips are slightly turned up. The somewhat fleshy **leaves** are scattered along its stems, but tend to be clustered towards their tips. They are quite aromatic, especially when bruised. The medium to deep green leaf blade is 1–2 cm long by 4–15 mm wide and has one to five pairs of pinnate lobes. The creamy **flower heads** are 8–16 mm in diameter, on stalks 5–10 cm long, each stalk

carrying a single, typically dome-like flower head. The flowers are quite strongly scented, the scent having a suggestion of honey. From a distance it is not unpleasant but is less pleasant close up or in quantity. Flowering usually occurs from December to February. *Leptinella pyrethrifolia* is found in subalpine to high-alpine regions of the North and South islands from the Ruahine Range to the Tararua Range in the North Island and is fairly wide-spread in most South Island mountain areas to as far south as Canterbury and Central Westland. It ranges from 600 to 2000 m. It generally occurs in moist, open sites such as stream-banks, gravel banks, shady rock ledges, damp grasslands, fellfields and herbfields.

South Island edelweiss *Leucogenes grandiceps*

New Zealand has four species of edelweiss, all of which are very charming, with two in particular being of quite noble appearance ('edelweiss' translates as noble or precious white). Superficially, they resemble the edelweiss of the northern hemisphere (which belong to a different genus, *Leontopodium*) and although no European will admit that the New Zealand species are just as charming, local opinion has it that they are a match for their northern hemisphere counterpart.

The South Island edelweiss is low-growing with prostrate **stems** that become more erect at their tips. Sometimes it is a relatively small plant while at others it will form quite large patches up to 60 cm or more across. Its silvery or white **leaves** are closely placed along its stems and they are spreading, but sometimes curving downwards. They are 5–10 mm long by 2–4 mm wide. The **flower heads** are 5–15 mm in diameter and are produced from the tips of the branchlets. Each head comprises a cluster of 5–15 smaller flower heads that are surrounded by up to 15 woolly bracts (modified leaves) and it is the woolly bracts surrounding these multiple flower heads that make these plants so appealing. Flowering is between November and March, although the flower heads will remain in good condition for quite a long time. The South Island edelweiss occurs in low- to high-alpine regions of the South and Stewart islands, ranging from 800 to 1900 m, being reasonably common throughout. It is mostly seen on rock outcrops, ledges and stable debris in fellfields.

North Island edelweiss *Leucogenes leontopodium*

This species has a greater resemblance to the northern hemisphere edelweiss than the other native species. In habit it is similar to the South Island edelweiss (see previous entry) and can truly be described as 'noble' or 'precious white', which is the translation of 'edelweiss'. It has a **woody rootstock** that can be much branched. Its growth is a bit more robust and the upper halves of its **stems** are upturned so that it can be up to 5 or 8 cm high. Its **leaves** are 8 mm–2 cm long by 4–5 mm wide and they are covered with tightly flattened fine hairs that give them a silvery or slightly golden appearance. It is with its **flower heads** that this species shows its full glory. They are similar to those of South Island edelweiss but are up to 3 cm or more in diameter. *Leucogenes leontopodium* occurs in low- to high-alpine regions of the North Island from Mt Hikurangi southwards through the central and southern mountains. In the South Island it is confined to the mountains of north–western Nelson and in the eastern Nelson–western Marlborough region, ranging from 1200 to 1800 m. Two recently recognised species (*Leucogenes neglecta* and *L. tarahaoa*) were until recently considered to belong to *L. leontopodium*. Both are similar but easily recognised. *Leucogenes neglecta* occurs in Marlborough on the mountains between the valleys of the Wairau and Awatere rivers. *Leucogenes tarahaoa* occurs on Mt Peel in Mid Canterbury, far to the south. The habitats for both of these species are similar to that of the North Island edelweiss, mostly rock outcrops, ledges and stable debris in fellfields.

Olearia nummulariifolia

A much-branched **shrub**, usually from 60 cm to about 2 m tall. It has rather stout yellowish or whitish **branchlets** which, when young, are often viscid or sticky, especially towards their growing tips. Its **leaves** are closely placed and either spreading to somewhat ascending. They are rounded, 5–10 mm long by 4–7 mm wide, and usually saddle-shaped. They are quite thick and leathery, their upper surfaces are bright green and shiny when young and their undersurfaces are white or buff-coloured. Its **flower heads** are white with a yellow centre, on stalks that may be longer than the leaves; normally they are solitary but occasionally there may be two or three. There are usually three to five ray florets (petals) to each flower head. Flowering generally takes place between November and April. *Olearia nummulariifolia* is found in scrublands and in mixed snow tussock–scrub in montane to subalpine districts of the North and South islands. It extends from the Raukumara Range and Lake Taupo southwards. Its altitudinal range is usually from 600 to 1370 m but in the southern part of its range it descends to lowland districts.

Mountain cottonwood *Ozothamnus vauvilliersii*

ASTERACEAE

The mountain cottonwood is a somewhat **aromatic shrub** up to 3 m tall, in more sheltered conditions, but may be less than a metre tall in alpine conditions. The **bark** of its main stems and branches peels off in small flakes. It has numerous small **leaves** that are 3–12 mm long by 2–3 mm wide, their upper surfaces are dark green while their undersurfaces have a dull, yellowish or yellowish-brown covering of thickly felted hairs. The usually white or creamy **flower heads** are 3–4 mm in diameter and are produced in rather tight clusters of 10–20 heads at the tips of the branchlets. Each head has several small, white-tipped, petal-like bracts (modified leaves). In some localities the outer scales of the heads are tinged with

red. It usually flowers between November and February and the spent flower heads remain on the bush for quite some time after flowering has finished. Mountain cottonwood occurs in the North, South and Stewart islands where it is found from East Cape and Lake Taupo southwards, ranging from sea level to 1500 m. It grows in open forest and scrub, subalpine scrub and mixed snow tussock–scrub. It is also known as tauhinu and mountain tauhinu. Formerly known as *Cassinia*, this species has reverted to its earlier name of *Ozothamnus*. The alpine form of this species is sometimes referred to as variety *montana*.

Vegetable sheep *Raoulia eximia*

Along with *Haastia pulvinaris* (also commonly known as vegetable sheep; see p. 83) this has to be one of the most extraordinary of our native plants. From a distance this **shrub** has a white or greyish white appearance that gives the impression of sheep being on the mountainside. *Raoulia eximia* forms large, irregularly shaped, cushion-like masses that are up to 60 cm high by about 2 m long and 1 m across. Its **branchlets** are so tightly packed that it is impossible to part them and as a consequence only the very tips of its branchlets are visible. The **leaves** have a tuft of white hairs on their tips so that their green or bluish green colour shows through in a rather ethereal manner. This creates a most beautiful effect that has been likened to viewing an icefall with its green light reflected through a thin veil of snow. As with the *Haastia*, the interior of the *Raoulia* **cushion** is a dense mass of branches and branchlets clad with the peaty remains of its old leaves. In winter it may be buried under snow for several months while during the summer it may be baked by the sun during the day and frozen by frost at night. The hairs on the outside of the cushion catch moisture that condenses from night fogs so enabling the plant to maintain its supply of moisture even under the driest of conditions. Its **flower heads** are about 3 mm in diameter and are sunk among the foliage at the tips of its branchlets. Flowering occurs during December and January. The use of the name 'vegetable sheep' for this species was first recorded in 1867, although no doubt it was used by high-country musterers well before then. The vegetable sheep occurs in high-alpine areas of the higher and drier mountains east of the main divide in the South Island, ranging from 1100 to 1800 m. It grows on shattered rock outcrops that protrude from shingle screes, in fellfields and on rock bluffs.

ASTERACEAE

This is an easily recognised species that is similar to *Raoulia subsericea* (see p. 96) but is readily distinguished by its brighter green colour and **looser mats** of growth which are usually up to 50 cm or more across. As its name *glabra* indicates it is almost completely hairless while the tips of its **branchlets** are usually more upturned (to about 2 cm) than the more prostrate tips of *R. subsericea*. Its narrow **leaves** are 3–6 mm long and are loosely placed along its stems. Their upper surfaces are light green or yellowish green and shiny. The creamy-white **flower heads** are up to 9 mm in diameter and do not have the papery, petal-like scales of *R. subsericea*. *Raoulia glabra* usually flowers between December and March. It occurs in the North, South and Stewart islands from Mt Taranaki and the Ruahine Range southwards, ranging from sea level to 1300 m. It usually grows in lowland to low-alpine areas where it may be found in grasslands and herbfields, on riverbeds, and in stony and rocky places and coastal and dune hollows. Some authors incorrectly refer to all mat-forming species of *Raoulia* as 'scab-weeds'; however, that name really applies only to *R. australis*, the notorious scab-weed of Central Otago and South Canterbury, which is found only in lowland to montane regions.

While this is one of the smaller species of *Raoulia* it has the largest flower heads of any species. It is quite distinct and, from its appearance, could easily be mistaken for a small species of *Celmisia*. It is a creeping plant that forms rather loose, **silvery patches** or **cushions** up to 15 cm or so across. Its **stems** are much-branched and they are clad with the remains of the old leaves. The pointed, silvery **leaves** are closely overlapping, narrow and about 5–10 mm

long. Its **flower heads** are produced from the tips of its stems; they are about 1.5 cm in diameter and are surrounded by white, papery, petal-like scales or bracts. It usually flowers between December

and January, although the flower heads are quite long-lasting and even when spent give the impression that the plant is still in flower. *Raoulia grandiflora* occurs in both the North and South islands in low- to high-alpine areas from Mt Hikurangi and central parts of the North Island southwards, ranging from 1000 to 1900 m. It is found in a wide variety of habitats and may be present in most types of alpine vegetation. It grows in snow tussock grassland, herbfields, and fellfields from exposed to rocky ridges.

ASTERACEAE

Although some people confuse this and *Raoulia glabra* (see p. 94) the two are really quite easily distinguished. *Raoulia subsericea* forms **compact mats** some 30–40 cm across whereas those of *R. glabra* are somewhat looser and usually of a lighter and brighter green. *Raoulia subsericea* is much-branched and its stems root into the ground. Its often ashen-coloured **leaves** are 3–6 mm long and are quite narrow. They are deep to medium green and on their undersurfaces are rather sparsely to densely clad with a thin covering of flattened silvery or golden hairs, which is another obvious point of difference between the two species. It is with the creamy-white **flower heads**, however, that *Raoulia subsericea* can be so readily distinguished from *R. glabra*. They are about 1 cm in diameter and each head is surrounded by white, papery bracts (modified leaves) that resemble petals. The flower heads are distinctly scented, particularly in warm, sunny conditions. The flower heads of *R. glabra* do not have those distinctive, papery bracts. Flowering usually takes place between December and March, and even when flowering is over the heads remain a conspicuous feature of the plant for some time. *Raoulia subsericea* occurs in montane to subalpine regions of the South Island except for the wettest mountain areas, ranging from 400 to 1500 m. It is common in grasslands and open places, particularly in the drier hill and mountain country.

Tutahuna *Raoulia tenuicaulis*

This is another of the mat-forming species of *Raoulia* and it is easily recognised by its wide-spreading, silvery-green mats. It is a prostrate, closely matted **herb**, forming soft to occasionally lax patches of silvery-green that can sometimes be very extensive. Its **much-branched stems** root into the ground as they grow; the branchlets are short and slightly ascending at their tips. The loose to closely overlapping or rather wide-spreading **leaves** are about 3–5 mm long, with pointed tips. Their bright green upper surfaces are clad with silvery hairs. Its small, creamy-white **flower heads** are about 2–3 mm in diameter but they are not particularly showy. Tutahuna occurs in the North and South islands where it can be widespread

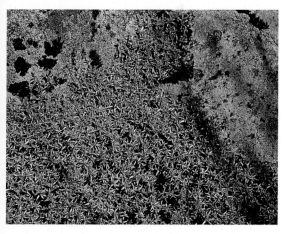

from about South Auckland southwards. It is quite abundant in lowland to low-alpine regions, ranging from sea level to 1500 m. It is often one of the most common mat plants, especially on stony or gravel riverbeds in the wetter regions of the South Island. In low-alpine areas it also grows in loose, fine debris on moist sites. In the southern North Island, tutahuna is also known as Tararua dishcloth, because trampers used to use its thick mats for cleaning their dishes.

GENTIANACEAE

One of the commoner species, this is represented by two varieties: variety *bellidifolia*, which mainly occurs in the North Island; and variety *australis* of the South Island mountains. It is a **tufted, perennial herb**, 10–15 cm tall, and has either a single or branched **rootstock**. This species may be recognised by its elliptic to spoon-shaped **basal leaves** that are crowded into rosette-like clumps. Its slightly fleshy **leaf blades** are 1–1.5 cm long by 5–7 mm wide and their upper surfaces are bright to deep green and shiny. The **flowering stems** are 4–12 cm tall and they may be branched or not branched. Its **flowers** are produced either singly from the tips of the stems or two to six flowers arise fairly close together in a flat-topped head. Each flower is 1.5–1.8 cm long and goblet-shaped. Variety *australis* is a stouter plant that often forms quite dense clumps 6–12 cm across. Its larger flowers are up to 2.5 cm long and are usually freely produced. Flowering usually occurs during January and February, although, depending on the season, flowering may extend until April. The New Zealand members formerly included in *Gentiana* have now been placed in the genus *Gentianella* as being distinct from the northern hemisphere species of *Gentiana*. *Gentianella bellidifolia* occurs in the North and South islands from Mt Hikurangi southwards. It is fairly widespread but may be often local in its occurrence. It is found in subalpine to high-alpine regions, ranging from 600 to 1800 m, usually in damp grasslands, herbfields and boggy places. In the South Island it is mainly high-alpine, occurring in the higher altitude grasslands and herbfields and the fellfield.

Eyebright *Euphrasia revoluta*

Eyebright is a low, tufted perennial **herb** that is usually up to about 5 cm tall and with a somewhat woody **rootstock**. Its slender **stems** may be partly trailing and become more erect towards their tips. The narrow **leaves** are 2–10 mm long and do not have any stalks. Towards their tips they have just one pair of small, sharp teeth just below the large triangular or rounded tip. It usually has one to few relatively large **flowers** at the top of each stem. The 1–1.5 cm long flowers are white with yellow in the throat. Flowering is usually between December and March. Euphrasias are semi-parasitic, usually attaching themselves to the roots of grasses and other plants. Whether all of the native species conform to that is not really known. A number

of *Euphrasia* species are only annual and grow afresh from seed, each year. Eyebright occurs in the North and South islands from the Ruahine and Tararua ranges southwards. It is found in low-alpine to high-alpine regions, ranging from 900 to 1700 m, and is quite widespread in the South Island. It grows in a wide variety of habitats and often may be plentiful in boggy and open places, tussock grasslands and herbfields.

This species of hebe is classified as a semi-whipcord hebe. It is a **low shrub**, of rather straggling growth, and forms patches up to 20 cm across by 10–20 cm tall. It is easily recognised by its **angular branchlets** having closely overlapping, deep to bright green leaves, rather widely spaced, and all pointing forwards towards the tip of the branchlet. The **leaves** are about 4 mm long by 1 mm wide and their upper portions are more or less rounded with swollen tips. Their margins have scattered, tooth-like hairs. The white **flowers** are produced on small, short inflorescences about 6 mm long that almost hide the tips of the branchlets. Flowering is normally from December

to January. *Hebe ciliolata* is confined to the South Island and occurs from Nelson southwards to at least as far as Otago. It grows in low- to high-alpine areas usually along or close to the main divide, ranging from 1000 to 2000 m. It most commonly grows on rock ledges, in clefts, and especially in fellfields it may grow on stable debris slopes. It may also occur in moraine and on exposed rocky sites in snow tussock–herbfield. One locality where it can be readily and easily seen is the Otira Valley at Arthur's Pass.

Hebe epacridea

Hebe epacridea has a **trailing or sprawling** habit. It can be easily recognised by its leafy stems having rather hard, little leaves that are joined at their bases, which tend to give the **branchlet** a distinctly square appearance. Usually, because of its sprawling habit it is no more than a few centimetres high but may be up to 40 cm across. Its rigid **leaves** are wide-spreading and are about 5–7 mm by 4–6 mm, and they are strongly keeled. Their widened bases are joined together so that they encircle the stem. The white **flowers** are produced in compact inflorescences around the tip of the branchlet. Flowering usually occurs from December to March. The arrangement of the leaves around the branchlets ensures that they are not overly exposed to the sun and other harmful elements. At any one time, no leaf is entirely exposed but is always afforded some protection by the surrounding leaves. The thickened margins of the leaves also provide further protection. *Hebe epacridea* is confined to the South Island from Nelson and Marlborough southwards to Southland. It is widespread in high-alpine areas about and east of the main divide but is more common on the drier eastern ranges. It ranges from 1200 to 2900 m and along with two other species of alpine plants ascends to the highest altitude of any native vascular plant. Usually, it grows on stable scree and debris slopes, rock outcrops and in fellfields.

This species has the largest flowers of any hebe and fully justifies its specific name of *macrantha*, which means 'large-flowered'. When up in the mountains, its large chalice-shaped flowers are a very fine sight. In the wild, it tends to be a rather **straggling shrub** 30–60 cm tall. Its branches are usually quite stout and erect. The broad, thick and leathery **leaves** are 1.2–2.5 cm long by 7–12 mm wide. Their upper surfaces are shiny, usually a pale to yellowish green, and their margins are distinctly toothed. Its large, white **flowers** are 2.5–3 cm wide and they are quite closely placed around the tips of the branchlets in groups of two to six. Flowering takes place between November and March. *Hebe macrantha* is confined to the South Island in subalpine to low-alpine regions. There are two varieties recognised: the typical form is found from the central Southern Alps to Fiordland. It is usually on grassy alpine slopes and in short, subalpine scrub on steep, rocky places. *Hebe macrantha* var. *brachyphylla* is distinguished by having more rounded and red-margined leaves. It occurs in the wetter mountains of the northern part of the South Island from Nelson and the Wairau area of Marlborough to the Amuri district of North Canterbury (see lighter shading on map). It usually grows in damp, often rocky sites in mixed snow tussock–scrub and herbfield. Both varieties range from 760 to 1500 m.

Hebe tetrasticha is a small **shrublet** that is usually about 20–30 cm across and no more than 20 cm high. As with *Hebe ciliolata* (see p. 100), it is classified as a semi-whipcord hebe. Its fine, squarish **branchlets** are about 2.5 mm in diameter. (Its specific name of *tetrasticha* means 'arranged in four rows' and refers to the shape of the branchlets.) The minute, deep green **leaves** are so densely crowded along the branchlets that it is not possible to thrust a pin between them, while its quite small white **flowers** are clustered around the branchlets so as to hide their tips. Unusually for a hebe, there are normally separate plants with male and female flowers. It usually flowers between November and January. *Hebe tetrasticha* is confined to the South Island where it grows in subalpine to high-alpine regions of the drier eastern ranges of the Canterbury mountains, ranging from 800 to 1800 m. It usually grows on rock faces and bluffs but occasionally it is also found on ledges and in loose, stony debris.

This is one of the smaller species of *Ourisia*, and it forms bright green **mats**, usually in damp places. When flowering it is a delightful little alpine. Its creeping **stems** root into the ground as it grows. The somewhat fleshy **leaves** are in pairs arranged in rows along either side of the stems, and they are 6–10 mm long, usually having one or two small notches on each side of the margin. Its **flowers** are on erect stalks 4–10 cm high with the flowers being mostly in pairs. The **corolla** is about 2 cm in diameter, with two lobes (the two uppermost) turned up and three turned down. Flowering mostly occurs between November and February. *Ourisia caespitosa* occurs in the North, South and Stewart islands where it is found in montane to high-alpine regions from Mt Hikurangi southwards, but is absent from Mt Taranaki. It ranges from 700 to 1800 m. It grows in a variety of habitats but is commonest in damp or shady, rocky sites, along stony stream-banks, and on bluffs and rock faces; it also occurs in fellfields and on shallow snow-banks.

Ourisia lactea is typically a South Island and Stewart Island species, except that one subspecies of it (subspecies *drucei*) does extend into the North Island (see lighter shading on map). When in flower it can be quite spectacular.

The South Island form (*Ourisia lactea* subspecies *lactea*) has rather thick, creeping and rooting **rhizomes** and at times may form quite large colonies. Its rather dark green **leaves** are 4–12 cm long by 3–7 cm wide, their undersurfaces often suffused with a purplish colour. The margins of the leaves are regularly and bluntly toothed. The stout **flower stems** are 20–45 cm tall and purplish, and they bear three to seven **whorls** of flowers. The white **flowers** have yellow in their throats and are on stalks 3–5 cm long while the individual flowers are 1.5–2.5 cm in diameter. Flowering can be between November and April. *Ourisia lactea* subspecies *lactea* occurs in montane to subalpine regions throughout the South and Stewart islands in all but the driest parts of eastern areas and Central Otago, and also occurs on Banks Peninsula. It ranges from 600 to 1500 m, and usually grows on rocky bluffs, along moist stream-banks, in rocky places in open subalpine shrubland, forest margins and damp banks.

Ourisia macrocarpa

PLANTAGINACEAE

This species is somewhat similar to *Ourisia lactea* subspecies *lactea* (see previous entry) but is usually larger and more robust and it lacks the hairs that characterise the leaves of that species. It may be recognised by its broad, almost rounded **leaves** that are a bright, shiny green and more or less heart-shaped at their bases. It is a stout, **perennial herb** with a **creeping rootstock** that may be 1 cm or more in thickness. The **leaf blade** is

Ourisia macrocarpa

4–12 cm long by 2–10 cm wide, quite thick and leathery, and the veins on its upper surface are indented while the margins are bluntly toothed. Its stout **flowering stem** is 25–50 cm tall and is rounded or slightly angled. The **flowers** are produced in whorls, five to nine per whorl, and they are on stalks up to 8 cm long. They are white with yellow in their throats and are 2–3 cm in diameter. Flowering occurs between November and February depending on altitude. *Ourisia macrocarpa* is the typical form (variety *macrocarpa*) which

occurs in subalpine to low-alpine regions of Fiordland, while *O. macrocarpa* var. *calycina* is the most widely distributed variety, being found from south-western Nelson to about central Westland

(see lighter shading on map). It is by far the most handsome of the ourisias. It is distinguished not only by its large size (the leaves being up to 15 cm by 10 cm) but mainly by the bases of its leaves tapering to the strongly purple petiole and not being heart-shaped as in the typical form. Its flowers are 2–4 cm in diameter. Both varieties occur in the higher rainfall regions, especially

Ourisia macrocarpa calycina

about and west of the main divide in subalpine shrublands, mixed snow tussock–herbfield, stream-sides and rock bluffs. The varieties range from 760 to 1370 m.

In some respects, this species is not dissimilar to *Ourisia macrocarpa* (see previous entry) but differs principally in being hairier. The **leaves** of *O. macrophylla* are at least as hairy beneath; other parts also have numerous hairs. It is a large and variable **herb** with a thick, **creeping rootstock** that terminates with a tuft or rosette of more or less erect leaves. There are two subspecies: subspecies *macrophylla*, the typical form, which is restricted to Mt Taranaki; and subspecies *robusta*, which occurs in the central and western North Island. The medium green **leaf blade** is around 4–10 cm in diameter and its upper surface is more or less furrowed because of the principal veins. The margins of the leaves are evenly and bluntly toothed. The base of the leaf varies from being slightly heart-shaped to being somewhat tapering to the stalk. Its stout **flowering stem** is up to 40–50 cm

Ourisia macrophylla robusta

tall and bears the flowers in up to eight whorls. The white flowers are 1.5–2 cm in diameter and have yellow in the throat while the backs of those of subspecies *robusta* are often flushed with purplish red. They are on stalks 3–5 cm long. Flowering occurs at any time between October and January, although it is more usual between November and December. Subspecies *macrophylla* is confined to Mt Taranaki and its adjacent Pouakai Range, from 800 to 1400 m altitude, and is common in sub-alpine scrub and low-alpine herbfield. Subspecies *robusta* occurs from the Raukumara Range southwards to the central North Island mountains, the Manawatu and eastern Taranaki (see lighter shading on map). It usually grows in montane to low-alpine regions, ranging from 100 to 1000 m, where it may be common in tussock–herbfields, alongside streams and on damp banks.

Parahebe hookeriana

PLANTAGINACEAE

This small species of *Parahebe* has a **trailing or creeping** habit and grows only a few centimetres in height while its much-branched stems may be up to 25 cm in length. The thick, leathery **leaves** are rounded and about 4–12 mm by 3–8 mm. Their upper surfaces are deep to medium green and their margins are deeply to shallowly toothed. The **flower stems** are about 2–6 cm long and they usually have only a few (four to eight) flowers on them. Its **flowers** are quite large, up to 1 cm in diameter, and are mostly lavender but may vary from white to different purplish and pinkish shades. There is an ocular ring inside the flower with magenta lines radiating outwards from it. Flowering usually occurs between about December and March. *Parahebe hookeriana* is confined to the subalpine to high-alpine regions of the North Island, where it extends from the Raukumara Range to the Kaimanawa Mountains, Huiarau Range, Maungaharuru Range, Ruahine Range and the Volcanic Plateau. It ranges from 900 to 1800 m. It usually grows in rocky places, especially on exposed ridges in tussock grassland–herbfield or in loose, stony debris in fellfields.

Parahebe lyallii

Parahebe lyallii is a small, creeping and rooting, much-branched **subshrub** that is usually no more than 5–7 cm high. It is a common species that is often found alongside streams and on stable riverbeds, occurring in many areas of the South Island. Its small **leaves** are 5–10 mm long and they have up to two or three pairs of small notches or teeth around their margins. The **flowering stems** are 3–8 cm long and carry numerous saucer-shaped **flowers** that are about 1 cm in diameter. The colour of the flowers is quite variable, from almost white to various pinkish shades, and each flower has a yellowish centre from which usually pinkish or purplish lines radiate. Flowering usually occurs between December and February.

According to locality, *Parahebe lyallii* can exhibit quite a variation of form and leaf shape and one or two distinct variations occur in the Nelson area. It is confined to lowland, montane and subalpine regions of the South Island, ascending to about 1300 m. Favoured habitats are alongside streams, on stable riverbanks, moraines, rock bluffs, snow tussock–shrub, herbfields and similar moist sites.

CAMPANULACEAE

The bluebell commonly comprises two similar species: *Wahlenbergia albomarginata* from the drier, eastern side of the Southern Alps; and *W. laxa*, which occurs in the higher rainfall regions about the main divide. Both are small **perennial herbs** with creeping, underground **stems** that produce small rosettes of leaves at their tips, although more often than not these rosettes are widely spaced and seldom form a complete mat. The **leaves** vary in size according to habitat. If the plant has adequate moisture, it will have larger leaves than those of a plant growing in impoverished conditions. They are 5 mm–4 cm long by 1–10 mm wide, being widest above their middle, and usually they are more or less spoon-shaped. The margins of *W. albomarginata* are usually smooth and whitish (as indicated by its specific name) while those of *W. laxa* are never whitish but have a few small teeth or are somewhat undulating. The **flowers** of both species are very similar, being borne on flowering stems 10–25 cm long, each bearing a single flower that tends to stand upright rather than hang downwards like a bell. Generally, they are 1.5–3 cm in diameter and vary from white to a pale or slightly more intense blue. Flowering can extend over quite a long period, which may be from November to February. *Wahlenbergia albomarginata* is found in lowland to low-alpine regions from eastern Marlborough and Canterbury to Central Otago and occurs mainly on grassy river terraces and in tussock grassland country. *Wahlenbergia laxa* is found in higher rainfall regions from north-western Nelson along the main divide to Westland, western Otago and Fiordland. It may be found in a variety of habitats from river flats and lake shores to subalpine rocks and ridges. Both ascend to about 1400 m.

Donatia novae-zelandiae

Donatia novae-zelandiae is a distinct and easily recognised species that forms hard **cushions** of deep green, particularly in alpine boggy places where the cushions may be quite extensive. In the wild, the cushions may be a metre or more across and, particularly in southern regions, the cushions often coalesce to form contiguous masses covering quite large areas. In fact, such areas are known as cushion bogs. The plant has short **stems**, buried in the cushion, and for much of their length they are covered with the remains of the old leaves which, in time, become quite peaty. Its **leaves** are densely overlapping, with only their upper halves free, and they are up to 10 mm long by about 1 mm thick. The apexes of the leaves are bluntly pointed. In midsummer the cushions become studded with star-shaped, white **flowers** about 4–6 mm in diameter that tend to be sunk among the leaves. Each flower has five, pointed petals. Flowering is usually between December and March. Visitors to the mountains often become confused with a similar-looking plant, *Phyllachne colensoi* (see p. 113), but the two are readily distinguished: *P. colensoi* has lighter yellowish green cushions, never the dark green of *Donatia novae-zelandiae*, and its leaves have very blunt points, in fact an almost cut-off appearance at their tips. Its flowers are similar to those of *D. novae-zelandiae* but have rounded, not pointed, tips to the petals. *Donatia novae-zelandiae* is found in the North, South and Stewart islands from the Tararua Range southwards, mainly in subalpine to low-alpine areas, ranging from 760 to 1520 m. It is found in cushion bogs overlying peat, permanently wet depressions in snow tussock or red tussock grasslands, and in herbfields.

STYLIDIACEAE

This is an attractive little alpine that may not always be particularly noticeable. It is almost a subshrub but lacks the woody bases to its stems. The **stems** are rather slender and up to 10–15 cm tall. The **leaves** may be fairly closely placed to somewhat more widely spaced apart. They are 6–12 mm long, narrowly oblong, with the upper surfaces deep green, slightly paler beneath and blunt at their apexes. Its **flower stalks** are up to 10 cm long and there are usually one to three flowers. The white **flowers** are 6–10 mm in diameter and have five to seven petals. Occasionally, the flowers of some plants may have a reddish or pink eye. Flowering usually takes place from about November to December. *Forstera bidwillii* occurs in the North and South islands from Mt Hikurangi near East Cape to the volcanic

mountains and then southwards. It is found in subalpine to high-alpine areas throughout but is frequently rare or absent from some areas in the South Island. It ranges from 800 to 1800 m, and usually occurs in wet, snow tussock–herbfield, in open scrub, on bluffs and similar rocky sites. This plant is named in honour of two botanists. The genus is named after one of the Forsters (father and son) who accompanied James Cook on his second voyage to New Zealand, and the species after John C. Bidwill, who was an early plant collector and explorer, especially active in the central volcanic region of the North Island and the Nelson mountains.

Phyllachne colensoi is a moss-like plant that forms hard and compact **cushions or mats**, usually of a medium to deep green or not infrequently a yellowish green. Much depends upon whether it is growing in a moist or dry situation. The cushion is usually 5–10 cm thick and mostly up to 20–50 cm across, but sometimes it can form more extensive cushions. Its **stems** are short and, as with *Donatia novae-zelandiae* (see p. 111), their lower portions are clad with the

old leaf remains, which eventually become quite peaty. Its **leaves** densely overlap and are about 4 mm long, tapering from a somewhat broad base to a very blunt tip that has an almost cut-off appearance. The white **flowers** are sunk among the leaves and just protrude sufficiently for the petals to spread open. They are distinguished from *Donatia novae-zelandiae* by their petals having rounded tips instead of being pointed and the fact that it sometimes has more than five petals to the flowers. *Phyllachne colensoi* is distributed in low- to high-alpine regions, throughout the North, South and Stewart islands, from Mt Hikurangi and the central volcanic mountains southwards, and ranges from 900 to 1900 m. It is often common in herb-moor overlying peat, in short, open, snow tussock–herbfield, also in snow hollows and on exposed ridges at higher levels, fellfield, cushion vegetation and shallow snow-banks. For a plant that often grows in boggy conditions it can be surprisingly tolerant of dry, exposed situations.

Lobelia roughii is another of the highly specialised and interesting plants that inhabit shingle screes. It is a **summer-green herb** with fleshy leaves and rather delicate **stems** that somehow manage to cope with the surface movement of the scree. It has numerous, **wide-spreading roots** that penetrate fairly deeply into the scree. When damaged its stems and leaves exude a **milky sap**. Its **leaves** are bronze to dark bronze and are a little more conspicuous than some scree plants. They are about 1–2 cm by 1–1.5 cm and their margins are coarsely and deeply toothed. The **flowers** are white to pale cream and are quite sweetly scented. They are usually produced on

stout stalks up to 5 cm long. Flowering generally occurs between October and January. After flowering, a large, purplish **seed capsule** develops. *Lobelia roughii* is found in the low- to high-alpine regions of the drier South Island mountains east of the main divide, ranging from 1000 to 1800 m. It is particularly confined to the greywacke shingle screes of the mountains from Marlborough and Nelson to north Otago. *Lobelia roughii* was named after its discoverer, a Captain Rough, who collected in the mountains of Nelson in the mid-19th century.

114

Panakenake *Pratia angulata*

Panakenake is a wide-spreading, **creeping and rooting herb** that can often cover quite a large area. The **stems** are quite slender and much interlaced. It is recognised by its rounded leaves and lobelia-like flowers. The **leaves** vary in size, generally being about 4–12 mm by 3–13 mm, and they have a few quite coarse teeth around their margins. The white **flowers** are produced on slender stalks 2–6 cm long, the individual flowers being 7 mm to 2 cm long; generally, the flowers of alpine forms are smaller than those from lowland areas. Flowering can be over quite a long period from October to April. The oval or rounded **fruits** are 7 mm to 2 cm in diameter, usually abundantly produced and are a magenta to purplish red colour. They may be present from January to July. Apart from anything else, this plant is easily recognised because of the way that its flowers are split along one side so that its five petals (corolla lobes) are unevenly spaced around it. As its fruits are often present at the same time as the flowers, they will also help to confirm its identity. Panakenake is distributed throughout the North, South and Stewart islands from sea level to low-alpine regions, around 1300 m. It is reasonably common in damp situations in subalpine grasslands, on banks, alongside streams, in open forest and herbfields.

Pratia macrodon is more compact and less wide-spreading than the related *Pratia angulata* (see previous entry). It is distinguished by its thicker, almost fleshier leaves, and its pale cream to creamy-yellow flowers. It forms small, rather **dense mats** or its growth may be a little more scattered. Its bright green, somewhat shiny, **leaves** are 5–10 mm long by 4–9 mm wide with coarse teeth around the upper half of their margins. The creamy-coloured **flowers** are almost without stalks and appear to sit on the foliage. They are also deliciously scented and particularly on a warm day their scent can pervade the surrounding air. Flowers from southern regions appear to be somewhat paler than those from the northern South Island. Flowering occurs between December and February. The **fruits** are greenish to slightly

purplish and are generally not very conspicuous, particularly as they nestle among the leaves. *Pratia macrodon* is fairly widespread in sub-alpine to high-alpine regions of the South Island from Nelson and Marlborough to Fiordland and southern Southland, ranging from 700 to 1900 m. It is found in grasslands, herbfields, rocky places, depleted grassland, shrublands, scree margins and fellfields.

BORAGINACEAE

Myosotis australis is a fairly common species of native forget-me-not and is mainly found in grasslands. The colour, referred to as forget-me-not blue, which most people imagine to be the typical colour of forget-me-not flowers is by no means typical and certainly does not apply to the New Zealand species. Of the more than 50 species of native forget-me-not only two, confined to the subantarctic islands, have blue flowers; the remainder all have white, yellow or bronze flowers. *Myosotis australis* is a **tufted herb**, forming rosettes up to about 7 cm tall, usually a short-lived perennial or biennial, and it can be recognised by its green leaves often being overlaid with a brownish colour, its yellow flowers and its dark, almost black flower stems. The medium to deep green **leaves** are 2–6 cm long by 4–12 mm wide, the upper surface often brownish green or suffused with brown, with the undersurface paler. The **flower stems** are few to numerous, quite dark to almost black and 15–20 cm long or even up to 30 cm. Its yellow **flowers** are about 6 mm long by 5–7 mm in diameter. Flowering usually takes place between October and February. *Myosotis australis* occurs in both the North and South islands in subalpine to low-alpine areas, ranging from 500 to 1500 m. It may be local in parts of the North Island such as around the Kaimanawa Mountains, while in the South Island it is common in Canterbury and Otago but more local elsewhere. The species may be found in tussock grasslands, tussock grassland–scrub and depleted areas of snow tussock grassland.

Forget-me-not *Myosotis macrantha*

BORAGINACEAE

Of the native forget-me-nots, this is probably the most handsome. It is quite a variable species, particularly as to flower colour, but all forms of it are most attractive. It is a **tufted perennial herb**, often forming one to several rosettes and, when not in flower, grows to 5–12 cm or so in height. Its rosette **leaves** are a medium to deep green, 3–12 cm long by 0.6–2 cm wide and covered with soft spreading hairs. The **flower stems** may be quite long, 10–15 cm but at times up to 30 cm. Its flowers are produced on what is often known as a **scorpioid inflorescence**, which means that it is curled like the tail of a scorpion and, as the flower buds develop, it slowly unfurls.

The inflorescence may be branched or not branched. The colour of the **flowers** varies from yellow to pale gold, brownish orange or bronze; they are up to 2 cm long and 6–12 mm in diameter. As well as being such lovely colours, they are also very sweetly scented. Flowering usually occurs between December and early February. *Myosotis macrantha* is found in subalpine to low-alpine regions in the wetter mountains of the South Island from Nelson to western Otago, ranging from 600 to 1500 m. It is usually confined to moist, rocky or stony sites such as damp ledges and crevices or rock faces, damp, shady ravines or rock gutters, wet scree or moraine, or in high fellfield.

Fairly common in alpine regions, this astelia is often quite a feature of the alpine vegetation: its large, often silvery tussocks usually stand out from afar. It is a variable species, particularly in size and leaf coloration and often each district has its own variations. It is a large, **tufted herbaceous plant** which usually forms strong clumps or tussocks up to about 80 cm tall. It has numerous **leaves** from 50 cm to 1.5 m long by 2–4 cm wide, that curve or arch out from the clump. Their colour varies from a light to medium green or sometimes a darker green to bronze colour and they have a semi-transparent covering of short hairs that often gives their surfaces a silvery appearance. Their undersurfaces are usually silvery-white or buff. The branched **inflorescences or flower heads** are concealed among the leaves so that they are not always particularly obvious. The male and female **flowers** are on separate plants. Flowering is usually between November and December. When the **fruits** of the female plant ripen they are orange to reddish-orange and are usually quite ornamental. They are eagerly devoured by native birds in the area and do not always last very long. The fruits usually ripen from about February or March onwards. *Astelia nervosa* occurs in the North, South and Stewart islands and is widespread in lowland to low-alpine areas of hill and mountain regions. In the North Island it is found from Mt Hikurangi to Lake Taupo and Mt Taranaki southwards. It is found throughout the South and Stewart islands. It usually grows in damp areas of mixed tussock grassland–scrub association, tussock grassland and herbfield. It is also found in open beech forest, and in Southland it occurs almost down to sea level where it grows in peat bogs. Ascends to 1500 m.

LILIACEAE

This is an attractive low-growing species of astelia that is distinguished by its leaves being relatively wide and stiff as well as being widely spreading. It usually forms **tufts** 10–30 cm high and not infrequently spreads into patches that may be up to 50 cm or more across. Its **leaves** are usually 10–30 cm by 1–2 cm and are characteristically spread sharply outwards from near their bases, which gives the tufts quite an open appearance. Depending on the district where they grow their colour varies and they may be green to silvery or bronze. Their **inflorescences** are small and sunk among the leaves so that they are rarely noticed. Their reddish orange **fruits** are similarly inconspicuous and are not very freely produced. *Astelia nivicola* occurs throughout the South Island, mainly in the higher rainfall alpine regions, ranging from 1000 to 1730 m. It is rather local in the northern part of its range but becomes more widespread in southern regions. It is also found on Mt Terako at the southern end of the Seaward Kaikoura Range. It usually grows in moist snow hollows, in snow tussock–herbfield and in mountain meadows.

Astelia petriei

Astelia petriei is an easily recognised species that is distinguished by its stout habit of growth and its stiff and leathery leaves that generally stand rather erect. It is a **densely tufted** plant up to 60 cm or so tall, and it generally forms quite **large clumps or colonies**. Its pale green and shiny **leaves** are 25–80 cm long by 2–3.5 cm wide, although occasionally they may be up to 6.5 cm wide. The upper surfaces of the leaves have whitish ribs or veins and their undersurfaces are silvery with quite distinct green veins. This species appears to rarely **flower** or its inflorescences are so well concealed among the leaves that they are rarely noticed. If it **fruits** they are up to 2 cm

long and are yellow-orange. *Astelia petriei* is confined to the South Island and occurs in the higher rainfall regions in low-alpine areas, mainly west of the main divide, from Nelson to Fiordland, ranging from 900 to 1520 m. In Canterbury it extends as far eastwards as Mt Peel. It usually grows in permanently wet, open sites, snow hollows, snow tussock–scrubland and herbfields.

This is one of the commonest of the six New Zealand species of *Bulbinella* and is easily recognised by its narrow, green or brownish green leaves and its smaller flower heads. It is usually deciduous, sprouting annually from a fleshy, tuberous **rootstock**. It is quite a strong-growing but slender plant, 30–50 cm tall (sometimes larger and up to a metre tall) that forms **single tufts** to **rather large clumps**. Its **leaves** may be erect to spreading, up to 45 cm long, but rarely more than 1.5 cm wide. The **flowering stem** is stout or slender, brownish and of variable length, depending on where it is growing. Its **raceme** is normally about 20 cm by 2.5 cm and usually overtops the leaves. The numerous, individual, bright yellow **flowers** are on short stalks about 1–1.5 cm long and each flower is up to 1 cm in diameter. Flowering occurs during November and December. The **seed capsules** are 5–7 mm long and soon turn black once they ripen. *Bulbinella angustifolia* is one species that has proliferated because of the burning off that has frequently occurred in the high country, its tuberous rootstock enabling it to survive the ravages of burning. Further, it is not palatable to browsing stock. The species occurs in eastern areas of the South Island from about the Hurunui River southwards, ranging from 500 to 1700 m. It is found in tussock grasslands, often in damp places, but it is also often locally abundant in the drier, depleted grasslands. Although commonly known as Maori onion, there is no evidence that the old-time Maori ever took sufficient notice of this plant to give it a name. Maori onion appears to have been an invention of an early-20th-century author and does not seem to have been recorded prior to 1906. It is also known as golden star lily.

Maori onion *Bulbinella hookeri*

Bulbinella hookeri is a most attractive species that adorns many an alpine mountainside. It may be recognised by its larger size and more robust habit. It differs from *Bulbinella angustifolia* (see previous entry) principally in its broader leaves. There are also differences between the seed capsules of the two species, with *B. hookeri* having a noticeably **larger capsule**. Generally, it will grow to about 30 or 60 cm tall but in some localities plants up to 1 m are not unusual. Its numerous **leaves** are up to 60 cm long and usually are 3 cm or more in width. Its **flowering stem** is rather stout and more or less triangular in section. The raceme is up to 40 cm by 5 cm and generally overtops the foliage. Its numerous golden-yellow **flowers** are about 1.5 cm in diameter and are on stalks 1.5–4 cm long. The **seed capsule** is 7–9 mm long. It usually flowers between November and January. *Bulbinella hookeri* is found in the North Island from Mt Taranaki and the Huiarau Range to parts of the Volcanic Plateau and the north-western

Ruahine Range. In the South Island it occurs from Nelson and Marlborough southwards to about the Lowry Peaks Range just north of the Hurunui River in North Canterbury. It ranges from 150 to 1520 m. The species can be very common in moist sites in tussock grassland and herbfields, especially in seepages and other damp places. In some areas it will form very extensive colonies. Another species, *Bulbinella gibbsii* var. *balanifera*, occurs in the Ruahine and Tararua ranges of the southern North Island and, in the South Island, about and west of the main divide from Arthur's Pass to Fiordland. As with the previous entry, although this species is commonly known as Maori onion, there is no evidence that the old-time Maori ever took sufficient notice of this plant to give it a name. The term Maori onion appears to have been an invention of an early-20th-century author and is actually quite misleading.

123

ORCHIDACEAE

Waireia stenopetala is an orchid that was formerly known as *Lyperanthus antarcticus*, an Australian genus, but it has now been elevated to the newly recognised genus of *Waireia*, which is confined to New Zealand and has just the one species. It is a small, upright orchid, between 10 cm and 30 cm in height. It has small underground **tubers**. The green **stem** is erect and slightly fleshy. It usually has two green **leaves** but occasionally there may be only one or as many as three. They are about 10 cm by 1.5 cm, generally spaced fairly well up the stem and sometimes they may have brown blotches. Usually, it may have one to four hooded **flowers** but there are more commonly only two and rarely exceeding three. They are about 1.5 cm long and are

green and normally have dull brown or reddish-brown blotches on their hood-like dorsal sepal that arches over the rest of the flower. Flowering is usually between December and February. *Waireia stenopetala* mostly occurs in subalpine to low-alpine regions of the North, South and Stewart islands, ranging from 800 to 1200 m. It also extends down to the subantarctic Auckland Islands. It normally inhabits bogs, wet, often peaty areas of tussock grassland, herbfield, open scrub and forest margins. It particularly favours alpine scrub and herbfields, especially if rather damp or wet.

124

Snow grass *Chionochloa pallens*

The large species of *Chionochloa*, or snow grass as they are commonly and collectively known, are a prominent and important feature of the lower alpine vegetation. Particularly in the wetter mountain regions snow grasses (tussocks) can be a dominant constituent of the mixed snow tussock–scrub and the snow tussock–herbfield associations, while in some regions they may also form a complete association, the snow tussock grassland, virtually by themselves. There are some nine different species of snow grass, as well as a number of subspecies. *Chionochloa pallens* is a quite common snow grass, also known as the midribbed snow tussock. It is a very handsome plant and fairly typical of the species of *Chionochloa*. It forms a stout and rather **tall tussock** up to about a metre tall and often has a pale or slightly yellowish appearance. Its **leaves** are variable and may be flattish, U-shaped or with its margins somewhat rolled inwards. One character that is used to help to distinguish the various species of snow grasses is the appearance of the **old leaf sheath**: whether the sheath remains whole or fractures into long segments or short segments, whether its margins remain separate or coil into a cylinder, or whether they are purplish or some shade of brown. The old sheaths of *Chionochloa pallens* do not fracture but remain intact. Its **flowering stems** are up to a metre or more long, but its **flower plume** is rather small and very open. *Chionochloa pallens* is found over quite a large part of both the North and South islands. It occurs in the higher mountains of the North Island from the Raukumara Range southwards. In the higher rainfall and western regions of the South Island, from northern Marlborough and Nelson southwards to Otago and Fiordland, it can be quite widespread in grasslands. It ranges from 1100 to 1800 m. As well as the typical form, *C. pallens* subspecies *pallens*, there are two other subspecies that occur in parts of its range.

Glossary

Alternate of leaves, or other organs, being placed singly along a stem or axis, not in opposite pairs.

Anther the pollen-bearing part of a stamen.

Apex the pointed end or tip of a leaf.

Axil the upper angle, usually between a leaf and a stem; adjective axillary.

Berry a fruit containing several to many seeds but not a stone.

Bloom a white or glaucous, waxy, powdery covering on some stems, leaves and fruits.

Bract a modified, often much reduced leaf, especially the small or scale-like leaves of an inflorescence.

Calyx the outer series of floral envelopes, usually green, each one referred to as a sepal.

Capsule a dry fruit which dries out and splits into several parts (valves) when mature.

Corolla the inner, usually showy whorl of floral parts, consisting of free or united petals.

Corymb a more or less flat-topped raceme with the long-stalked outer flowers opening first; adjective corymbose.

Deciduous a plant that sheds all of its leaves or the top growth of which dies down to below ground level in the autumn.

Dorsal relating to the back.

Elliptic (or **elliptical**) having the shape of an ellipse.

Floret a small flower, especially of the individual flowers in a composite head such as that of a daisy. The individuals around the outside of such a head may have a single, strap-like petal and be referred to as ray florets, while those in the central area that have no petals are referred to as disc florets.

Glabrous hairless, smooth.

Gland a minute organ that secretes oil, resin or other liquid, usually on the leaves but may also be on stems and flowers.

Glaucous having a whitish or greyish appearance, but not necessarily being due to a waxy or powdery bloom.

Globose rounded or ball-shaped.

Greywacke any dark sandstone or grit having a matrix of clay minerals; much of the rock along the eastern side of the Southern Alps is greywacke.

Inflorescence a general term for a collection of the flowering parts of a plant, or for the arrangement of the flowers.

Insolation the quantity of solar radiation falling upon a body or thing.

Local confined to individual sites within a locality or area and not widely distributed over a larger area.

Margin the edge or boundary line of an organ, particularly of a leaf but also of the parts of flowers.

Montane of or inhabiting mountain regions. In this work referring to those areas below the subalpine and alpine regions.

Obovate inversely ovate or egg-shaped with the broadest part towards the tip.

Ocular of or related to the eye, in reference to a flower when it has an eye-like ring of colour near its centre.

Opposite particularly of leaves, a pair of organs arising at the same level on opposite sides of the stem.

Ovate shaped like the longitudinal section of an egg, the broadest part being towards the base.

Panicle a loose, irregularly branched inflorescence, usually containing many flowers.

Parted usually expressed as '-parted'; referring to the number of parts of leaves or flowers, e.g. 'five-parted' = having five parts.

Pedicel a stalk supporting a single flower in a compound inflorescence.

Perfect a flower having both male and female elements present, both of which are functional.

Perianth the floral envelopes, either the calyx or corolla, or both; used particularly for flowers when the calyx and corolla are not well differentiated in form or one is absent.

Petiole the main stalk of a leaf; adjective petiolate.

Pinna a division, especially a primary division, of a divided frond (or leaf). Usually expressed as primary, secondary or tertiary according to the number of times a frond is divided; plural pinnae.

Pinnate a compound leaf (one with several leaflets) with the parts or segments arranged along either side of an axis, or midrib, as in a feather.

Pungent terminating in a hard sharp point.

Raceme an inflorescence having several to many, stalked flowers arranged along a single stem; adjective racemose.

Receptacle the enlarged uppermost part of the flower stalk on which the floral parts are borne.

Rhizome an underground stem, usually spreading more or less horizontally (creeping), or short and erect, and sometimes extending above the ground to form a short, erect trunk.

Sepal each of the divisions or leaves of the calyx.

Serrate applied to a leaf that has teeth like those of a saw on its margin.

Simple of leaves: in one piece and not being divided into leaflets like those of a compound leaf.

Sorus a cluster of sporangia (spore-containing structures) prominent on the fronds of most ferns; plural sori.

Spinescent having spines, or resembling spines, or becoming spiny.

Sporangium a sac or capsule that contains spores; plural sporangia.

Stamen the pollen-bearing organ of a flower comprising the anther and its supporting stalk or filament.

Stipe the stalk from which a frond blade is produced; plural stipes or stipites.

Strobilus fruiting cone; plural strobili.

Subacute almost or not quite sharply pointed.

Subalpine the lower parts of the alpine zone, above the tree line but below the true alpine zone containing herbfields, fellfields etc.

Subspecies a level just below that of specific rank and above that of variety.

Summer-green having its growth during the summer and dying down during the winter.

Talus a sloping, scree-like mass at the foot of a cliff. Usually, but not always, composed of larger fragments than a typical shingle scree.

Tepal an individual member of the perianth.

Terminal borne at the end of a stem and thus limiting its growth.

Tomentum densely matted, short, woolly, soft hairs.

Trifoliate having three leaves. Diminutive trifoliolate: having three leaflets.

Umbel a cluster of individual flowers where several flower stalks arise from the same point.

Further reading

Allan, H.H. 1961, *Flora of New Zealand*, Vol. 1, Government Printer.

Cave, Y. & Paddison, V., 1999, *The Gardener's Encyclopaedia of New Zealand Native Plants*, Godwit.

Metcalf, Lawrie, 1993, *The Cultivation of New Zealand Plants*, Godwit.

Moore, L.B. & Edgar, E., 1970, *Flora of New Zealand*, Vol. 2. Government Printer.

Salmon, J.T., 1968, *A Field Guide to the Alpine Plants of New Zealand*, A.H. & A.W. Reed.

Index

Notes

Notes

Notes

Notes

Other titles in the Photographic Guide Series:

Birds
978 1 77694 009 7

Ferns
978 1 77694 039 4

Fossils
978 1 77694 002 8

Freshwater Fishes
978 1 86966 386 5

Insects
978 1 77694 047 9

Mammals
978 1 86966 202 8

Moths & Butterflies
978 1 77694 003 5

Mushrooms & Fungi
978 1 990003 76 9

Reptiles & Amphibians
978 1 86966 203 5

Rocks & Minerals
978 1 990003 84 4

Sea Fishes
978 1 77694 052 3

Seashells
978 1 77694 053 0

Spiders
978 1 77694 041 7

Trees
978 1 77694 010 3

Wildflowers
978 1 77694 042 4